Thinking Like a Watershed

For Glenn Jatch
Remembering wonderful
days —
Jack Loeffler
2016

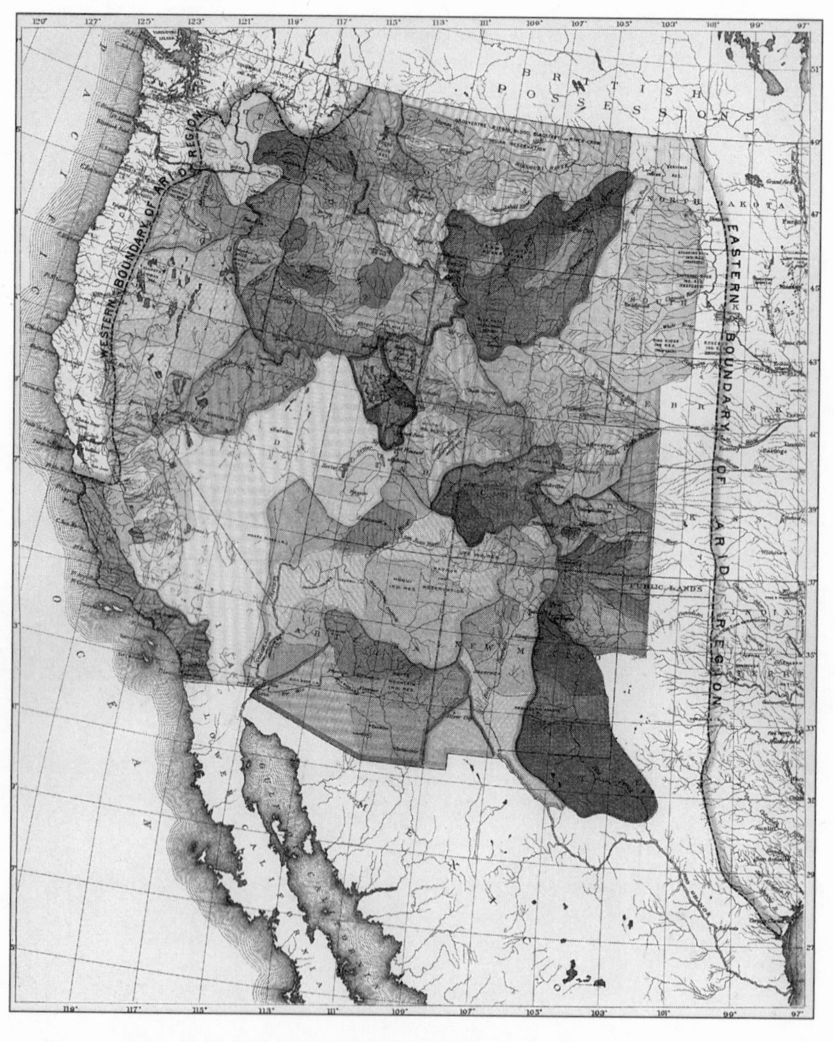

Arid region of the United States showing drainage districts, entire.
From *Eleventh Annual Report of the United States Geological Survey, 1889–90.*
Courtesy of the New Mexico History Library, Santa Fe

Thinking Like a Watershed

Voices from the West

EDITED BY JACK LOEFFLER
AND CELESTIA LOEFFLER

University of New Mexico Press ✢ Albuquerque

Printed in the United States of America
17 16 15 14 13 12 1 2 3 4 5 6

Library of Congress Cataloging-in-Publication Data

Thinking like a watershed : voices from the West / edited by Jack Loeffler and Celestia Loeffler.
p. cm.
Includes bibliographical references.
ISBN 978-0-8263-5233-0 (pbk. : alk. paper) — ISBN 978-0-8263-5234-7 (electronic)
1. Human ecology—West (U.S.) 2. Desert ecology—West (U.S.) 3. Stream ecology—West (U.S.) 4. Indigenous peoples—Ecology—West (U.S.) 5. West (U.S.)—Environmental conditions. I. Loeffler, Jack, 1936– II. Loeffler, Celestia, 1981–
GF504.W35T55 2012
304.20978—dc23

2012024090

BOOK DESIGN
Typeset by Lila Sanchez
Text composed in 10.25/13.5 Minion Pro Regular
Display type is ITC Fenice Std

For our children . . . may you remember the wisdom of the land

Contents

Acknowledgments

THE WATERSHEDS AS COMMONS PROJECT IS DIRECTLY DESCENDED FROM an earlier Lore of the Land project that included a fifteen-part documentary radio series entitled *The Lore of the Land* and a book entitled *Survival Along the Continental Divide*, published by the University of New Mexico Press in 2008, both funded by the Ford Foundation and the New Mexico Humanities Council.

The current project involves production of a fourteen-part documentary radio series entitled *Watersheds as Commons* and this anthology entitled *Thinking Like a Watershed*, which together comprise a massive aural history with transcriptions of over one hundred recorded interviews, the creation of the Lore of the Land website, a lecture series, and a one-hour radio documentary entitled *Aldo Leopold in the Southwest*. Jack Loeffler has been the project director for both, working closely with coproducer Celestia Loeffler.

This project could not have been accomplished without the extraordinarily generous funding provided by the Christensen Fund of San Francisco and the New Mexico Humanities Council of Albuquerque. We have been urged ever onward by Dr. Laurie Monti and Dr. Ken Wilson of the Christensen Fund and Dr. Craig Newbill, director of the New Mexico Humanities Council. They are true friends of Lore of the Land who believe, as do we, that biodiversity and cultural diversity are inextricably interlinked and that cognitive diversity is vital for conscious and conscientious human endeavors in the coming decades.

We are also grateful to our fellow board members of the Lore of the Land, including James McGrath, Enrique Lamadrid, Lynn Udall, Suzanne Jamison, Rina Swentzell, and Sue Sturtevant, and to our late founder, Lee Udall, and her late husband and Lore of the Land fellow board member, Stewart Udall.

We've had many years of good, solid common conversation with all of the above as well as with Bill deBuys, Gary Snyder, Bill Brown, Gary Paul Nabhan, Melissa Savage, Claude Stephenson, Richard Grow, Ed Abbey, David Menongye, and Alvin Josephy, all of whom contributed enormous consciousness to this project.

We're deeply indebted to the scores of interviewees without whose recorded words of wisdom our project would not have been realized.

We're especially grateful to our indefatigable, peerless transcriber and dear friend Yvonne Bond, without whom we'd be adrift in a sea of words.

Thank you to Lesley and Bob, who shared their time, labor, love, and their precious plot of land with Celestia for the growing season.

With humility and gratitude, we thank our families, who continue to be deep wells of moral support, honesty, and love.

And most of all, we are grateful to the landscapes, the watersheds, and the traditional peoples of the North American Southwest for revealing the spirit of place and for resisting secularization of habitat, no matter how hard we try to turn our homeland into money.

Thank you.

—Jack Loeffler and Celestia Loeffler
for the Lore of the Land

Preface

THINKING LIKE A WATERSHED IS AN UNCOMMON BOOK, PERHAPS A UNIQUE book. It is an anthology of points of view expressed by people of distinct cultural backgrounds, all of whom are profoundly imbued with the spirit of place that dominates the American Southwest. The Southwest is the most arid region of North America wherein water is the rarest of the four elements. This diverse landscape has nurtured many cultures for many millennia, each culture having evolved within the context of its respective homeland. Indeed, the environment has had an enormous influence on the shaping of every culture.

It is thought that humans have inhabited this landscape for at least twelve thousand years when megafauna including wooly mammoths, dire wolves, and even horses hunted and foraged through the last Ice Age that waned into eternity ten or so millennia past. The landscape has a long memory that is revealed in striations uncovered by water coursing through canyons, or by tectonic nudges that cause mountains to rise and rifts to sunder the land, by fossils to signal the presence of the long dead, and by artifacts and rock art to indicate the skill and imagination of ancient humans whose minds were aligned with the wild. Trees that have stood for a century mark the passage of the years in incremental rings that span cross-sections of tree trunks that indicate wet years and dry years, and that are able to be cross-referenced with other tree trunks extending hundreds of years into the past. Our species has the consciousness to interpret the land's memories and, one hopes, the collective conscience to work with the land in balance and harmony.

Several major cultural points of view are expressed in the pages that follow. In a way, each point of view is a window into the memory of the land. These points of view are expressed by members of Tewa, Tohono O'odham, Hopi, Navajo, Hispano, and Anglo cultures. Most of the authors were born in the American Southwest, and all of the authors have spent a great deal of time meandering through the landscape observing what takes place, ruminating on interconnectedness, and celebrating the spirit of place. As the great cultural anthropologist Edward T. "Ned" Hall pointed out, cultural and cognitive diversity are among humanity's greatest assets and provide myriad means of perceiving solutions to multiple problems—such as we face in the coming decades.

Six of the essays herein were written specifically for this anthology. The remaining five appear as excerpted interviews recorded and edited by Jack Loeffler and Celestia Loeffler. These "spoken essays" remain true to the manners of speech of the interviewees and reveal great depth of respective cultural perspective. The watersheds of this arid and exquisitely beautiful span of landscape are each distinct living bioregions whose myriad denizens live in a state of symbiosis with homeland. The writers and speakers represented in this anthology provide a compelling glimpse into the role of cognitive diversity as we proceed into an uncertain future.

Some of the inspiration for *Thinking Like a Watershed* came from the thinking of John Wesley Powell who forwarded the notion that the arid lands west of the 100th meridian should be seen as a mosaic of watershed commonwealths governed largely from within. The spirit of Aldo Leopold, who recognized the need for a land ethic guided largely by conscience, wends through parts of this book. However, the spirit of the mythic landscape of the American Southwest as revealed in the pages that follow is the true inspiration for this book.

Jack Loeffler, photo by Katherine Loeffler

Jack Loeffler

Jack Loeffler, Lore of the Land board member, is a bioregional aural historian, radio producer, writer, sound collage artist, and former musician. Since 1964, he has conducted field recordings west of the 100th meridian, founding the Peregrine Arts Sound Archive in 1967 to be the repository for his professional work, which he has donated to the New Mexico History Museum Library. His archive now holds thousands of hours of recordings of interviews, natural habitat, and over three thousand songs of indigenous and traditional peoples. His primary concern is restoration and preservation of habitat focusing on the relationships of indigenous cultures to respective habitats and the role of cultural diversity in attempting to solve the dilemmas now facing humankind. Loeffler has produced hundreds of documentary radio programs, as well as scores of soundtracks, albums of diverse musical genres, films, videos, folk music festivals, museum sound collages, and books. Selected radio series include *La Musica de los Viejitos*, *Southwest Sound Collage*, *Moving Waters—The Colorado River and the West*, *The Lore of the Land*, and *Watersheds as Commons*, produced with Celestia Loeffler. He has authored several books, including *Headed Upstream: Interviews with Iconoclasts*; *La Musica de los Viejitos*, including a three-CD set of Hispano folk music, with Katherine Loeffler and Enrique Lamadrid; *Adventures with Ed: A Portrait of Abbey*; *Survival Along the Continental Divide: An Anthology of Interviews*; and *Healing the West: Voices of Culture and Habitat*. He is the recipient of numerous awards, including a New Mexico Governor's Award for Excellence in the Arts, an Edgar Lee Hewett Award from the New Mexico Historical Society, an Archie Green Public Folklore Advocacy Award from the American Folklore Society, and a Stewart L. Udall Award for Conservation from the Santa Fe Conservation Trust. In the introduction to this volume he refers to many ways of looking at homeland within the context of watersheds and advocates for grassroots governance from within home watershed.

Thinking Like a Watershed

Introduction

"We give thanks to Evañu, the spirit of the river, for sustaining us," said the Tewa elder.

"Those Indians, when it comes to really understanding Nature, they know a hell of a lot more about that than the rest of us," said the old gringo. "They pray to it, and sing to it, and dance to it, and they've been here for a long time. Us, we just use it up."

"The acequias were our commons," said the Hispano elder. "Everybody got to share the water so we could irrigate our crops. During the growing season, everybody had their own fields. But after the harvest, everybody's cattle and sheep could graze in anyone's fields. Nobody privatized water till the Anglos came, and they made up their own laws. How can you sell water rights? The water belongs to everyone. It belongs to itself."

To many, if not most of us in America, water comes from the spigot or the showerhead or the commode behind the toilet just as electricity comes out of sockets in the wall. We are plumbed and wired within our immediate environments with little or no thought to Earth's sustaining grace. For five generations, most urban Americans and many rural folks have luxuriated in instant water and instant electricity with but a vague notion of the true source. Fortunately, cultural consciousness is beginning to register that problems may well lie ahead, and to solve these problems we may have to shed biases that have been with our ancestors for centuries.

As but one example, our species may be so genetically wired to "go forth and fructify" that we've absorbed that urge into our deepest cultural tenets to the extent that to many, birth control is a sin. From the point of view of the rational thinker, considering that the human population of the planet nearly quadrupled during the twentieth century while hundreds of other species of biota went irrevocably extinct and natural resources dwindled in direct proportion to human population growth, it is obvious that we as a species are fructifying out of control and that unless we lessen our numbers, we'll crash, perhaps to follow the dodo into oblivion. To fail to heed this realization is not only irrational, it's tragic. To think that a species blessed with our level of capacity for consciousness continues to dance blithely ever closer to the edge of the abyss, driven by an array of archaic tenets spiced with obsolete paradigms, is absurd. Yet such seems the case.

Some scientists now recognize that societies can respond as a single "superorganism" to selective pressures. Apparently, responding to rationality alone is insufficient to sway world society into balance. We must tap deeper into the collective human psyche to uncover regions to nurture. Compassion and conscience hopefully exist in virtually every human and increase the breadth and depth of individual and collective purview. Compassion and conscience are fundamental to any system of ethics that can be truly heartfelt, individually or within a human culture. Compassion and conscience may well waver in the face of desperation to survive, and indeed much of the human population of the world is now dangerously desperate.

But we have to start somewhere, so let's start in the American Southwest, where aridity is the main characteristic, and human communities, albeit growing too rapidly and excessively, still exist in relative proportion to carrying capacity of respective habitats and watersheds. At the same time, we should identify at least part of the array of collective human conduct modes through which we threaten not only ourselves, and societies everywhere, but also our fellow biota and even geophysical characteristics that have made our tenure as the keystone species possible.

In 1968, human ecologist Garrett Hardin wrote his evocative essay, "The Tragedy of the Commons," wherein he illustrated by historic example in Britain that when the carrying capacity of the commons grows too great, "the whole thing is destroyed."[1] Hardin had hearkened to the common sense reflected in "An Essay on the Principle of Population"[2] by the Reverend Thomas Robert Malthus published 170 years earlier. In 1972, a group of scientists at Massachusetts Institute of Technology published a book entitled *Limits to Growth*, wherein they provided models of the effect of a rapidly growing human population in a world of finite resources. They have updated their book twice, and many reviewers have taken great umbrage at their depressing audacity. Yet their presentations are honest reflections of their interpretations of the exponential increase of five major elements—population, food production, industrialization, pollution, and consumption of nonrenewable resources—and their interactions.

"Extrapolation of present trends is a time-honored way of looking into the future, especially the very near future, and especially if the quantity being considered is not much influenced by other trends that are occurring elsewhere in the system. Of course, none of the five factors we are examining here is independent . . . Each interacts constantly with all the others. Population cannot grow without food, food production is increased by growth of capital, more capital requires more resources, discarded resources become pollution, pollution interferes with the growth of both population and food."[3]

Notably, other factors such as pandemics, world war, and global warming could be, but as yet have not been, included in this model. In part, the point of the drill is to conceive all of the factors simultaneously and in constant motion. This requires a sophisticated form of nonlinear thinking that may only be possible for most of us with a computer. However, it's not that hard to memorize the five factors and then imagine them interrelating on different levels. Again, they are the following:

population
food production
industrialization
pollution
consumption of nonrenewable resources

This is a handy sphere of reference as we wander into the American Southwest, enchanted by the relatively clear light and sense of vast space, quietude, and peace.

Cultures blossom and wither with the passage of millennia. Six hundred years ago, nearly a century before European cultures were scheming on empire building in the new world, desert people of the Hopi village of Oraibi hunted, gathered, and practiced a form of agriculture as they already had for centuries. Today, they continue to participate in an annual cycle of ceremonials wherein they collaborate with spirit beings in the practice of survival, high-desert style. Six centuries ago, the Hopis were not alone in what is now regarded as the American Southwest. Other pueblos existed along the Río Grande; the pueblo known as Zuni thrived in the high country on the western aspect of the Continental Divide; nomadic Athabascan peoples—Apaches and Navajos—roamed the region, sometimes raiding Puebloans and other cultures; Ute Indians lived in proximity to the San Juan River; and far to the south, O'odham cultures eked out existence in what is now called the Sonoran Desert, where earlier cultures are known to have preceded the O'odham by more than ten thousand years. All of these culturally integrated humans were equipped with generations of lore, both practical and mythic, that recounted states of mind and empirical methods of practice appropriate for their respective homelands. Many of their descendants continue to regard fellow biota as kindred, and it is that refined sense of empathy for all life within this sacred landscape that defines their role in an otherwise frenetic techno-economic global culture.

Over four centuries ago, Spanish colonists questing for gold and human souls to convert to Christianity trailed along the Río Grande and began to settle the northern reaches of what is now New Mexico. Their first century was marked in great measure by mutual antagonism between themselves and their Native neighbors. The colonists brought with them missionaries representing a transcendental Christian god that would try without success to subsume the local deities, although the Christian mantle would come to share space with the katchinas. The colonists were violently expelled by Native peoples of different cultural persuasions in 1680 in a remarkably well-coordinated revolution, now recalled as the Pueblo Revolt, led by a Puebloan of extraordinary character named Popé.

A dozen years passed, and the colonists returned and over time learned to live with their Native neighbors in a state of mutual cooperation and intercultural exchange of both custom and blood known as the *mestizaje*, or mixture. The colonists tapped deep roots into the high-desert country and gradually established their own sense of indigeneity, much of their history recorded in narrative ballads and folk dramas that serve as mnemonic means of cultural recollection. They created great systems of acequias, or irrigation canals, a custom shared with neighboring Puebloans who themselves had created stone constructions to capture water, returning some to the aquifers, and otherwise expanding riparian habitat, fully aware of the water cycle that included clouds that stored rain, hail, and snow.

Across the span of the Great Plains to the east, Anglo culture, spawned ancestrally in Western Europe, was celebrating newly won liberation from the British Empire. Between 1804 and 1806, Meriwether Lewis and William Clark led an expedition to the coast of the Pacific Northwest, taking measure of the recent Louisiana Purchase. Their return heralded recognition of the immanence of westward expansion sparked by the youthful exuberance of a new nation, the United States of America. In 1822, an adventurous entrepreneur, William Becknell, followed trails west from Missouri to Santa Fe, a community founded in the first decade of the seventeenth century and capital of the northern reaches of the new Republic of Mexico. Becknell established the Santa Fe Trail as an important nineteenth-century trade route that saw transport of the highly prized fruits of the Industrial Revolution from the east in return for silver coins, woven goods, and raw materials from the west. Much of this occurred to the dismay of Plains Indians mounted on horses acquired from Hispanos, Native Americans who rightly reckoned that their cultural days were numbered as technologically superior Anglos infiltrated their hunting grounds, ultimately committing buffalocide on bison that roamed the plains as an integral part of the biotic community, animals that were traditionally hunted by Indians of many cultures for food, clothing, bone tools, and immense spiritual well-being.

As the eminent author Paul Horgan pointed out, "Of course, the very first motive was commercial, the coming of the Anglos. And though not a wholly ignoble motive, it certainly was a selfish one. Therefore, something of that emotional commitment to a purpose had an enduring effect on all relationships between the occupants—namely, the Indians, and the Hispanos and the incoming Yankees, Anglos. I know that superior judgements were rendered upon the inhabitants by those who came."[4] Horgan also provided insight into the vast cultural differences that existed between the Indians and the Hispanos and Anglos and that abiding sense of alienness that greatly colors perspectives of all cultures in collision.

Journalist Jonathon O'Sullivan expressed Yankee exuberance when he coined the term *Manifest Destiny*, as he wrote that it is "by the right of our [America's] manifest destiny to overspread and to possess the whole of the continent which Providence has given us for the development of the great experiment of liberty and federated self-government entrusted to us."[5] At the time, America was involved in a boundary dispute with Great Britain in the Oregon Country. O'Sullivan had earlier used the term regarding U.S. annexation of Texas. Thus he insinuated the notion that it was America's God-given right to take over the land. The term caught on and ostensibly influenced President James Knox Polk and thus contributed to the kindling of the Mexican-American War that ended in 1848. The war's end resulted in the United States winning the New Mexico Territory away from Mexico, to whom the United States paid $15 million plus a bit of change to settle some old scores to purchase moral righteousness along with the deed to a swath of land that extended through the American Southwest all the way to the Pacific Ocean. The war officially ended with the signing of the Treaty of Guadalupe Hidalgo.

In 1853, the United States considerably expanded the newly won landscape with the Gadsden Purchase that included nearly thirty thousand square miles extending from La Mesilla, New Mexico, to Yuma, Arizona. The United States of America was now an empire that spanned the continent from the Atlantic Ocean to the Pacific.

WESTWARD EXPANSION FIRED CULTURAL CONSCIOUSNESS. AFTER THE Civil War ended in 1865, a steady stream of homesteaders, entrepreneurs of every ilk, lawless bands of gunslingers, and the U.S. Army invaded the West, displacing Native Americans, especially Apaches and Navajos who bravely defended their homelands, many to the death. Some have regarded New Mexico Territory to have been the most outlaw-ridden, violent region in America during the latter half of the nineteenth century.

It was into this milieu that a one-armed adventurer, explorer, scientist, and endlessly curious seeker of knowledge—John Wesley Powell—began his explorations of the American West. In 1869, he and his small coterie descended the Green and Colorado rivers in wooden dories, being the first, as far as is known, to have ever attempted such an adventure. Born in 1834 in New York State, his family gradually settling in Boone County, Illinois, Powell ran part of the Mississippi River alone as a youngster, his vigorous curiosity regarding natural history his dominant characteristic. He lost most of his right arm to a minié ball wound at the battle of Shiloh where he served the Union forces as an officer. He felt pain at the end of the stump of his arm for the rest of his life but never allowed himself to be daunted.

He rode through much of the American West unarmed, befriending Native Americans as he went, even learning as much as he could of their languages and customs. He became the first director of the Bureau of American Ethnology, a position he held until his death in 1902. He was also the second director of the U.S. Geographical Survey from 1881 until 1894. It was during this period that he wrote about the American West that he perceived with far greater clarity than many if not all of his peers. He saw that the lands west of the 100th meridian were arid lands that received far less annual precipitation than lands in the verdant East. The Rocky Mountains separated the Great Plains from the Intermountain West, that expanse home to the Columbia Plateau, the Colorado Plateau, and the Basin and Range Province and bounded on the west by the Sierra-Cascade mountain ranges.

The Intermountain West is drained to the sea by only three river systems: the Columbia-Snake Watershed that drains to the Pacific; the Colorado River Watershed that drains to the Sea of Cortez; and the Río Grande Watershed that drains to the Gulf of Mexico. The Continental Divide runs roughly north to south along the Rockies, separating drainage systems east from west. The Basin and Range Province is home to the four deserts of North America: the Great Basin Desert of Nevada and part of Utah; the Mohave Desert of southern California; the Sonoran Desert, the most luxuriant, of southern Arizona and northwestern Mexico; and the largest, the Chihuahuan Desert of New Mexico, Texas, and part of northern Mexico. The Colorado Plateau contains the most intricate system of canyons in the world. The sun shines in New Mexico for more than three hundred days a year, illuminating the land of clear light.

Powell reasoned that easterners venturing westward were destined for hard times due to aridity. Farming would have to rely on irrigation rather than rainfall. And settlers would have to rely on farming and ranching to subsist. Gradually, Powell envisioned that the arid West be perceived as a mosaic of watersheds and that as westward expansion proceeded, each watershed would be settled, its waters to be shared for agriculture within the watershed, its timber and pastures to provide bounty for the settlers. This was admittedly an anthropocentric approach, part of the heritage of Manifest Destiny. Yet had Powell's dream of watersheds as commonwealths governed largely from within rather than without been approved by Congress, human cultural geography could well have been more deeply imbued with the realization of the flow of Nature with much less emphasis on economics.

On Saturday, March 15, 1890, Major John Wesley Powell addressed the Select Committee on Irrigation of the U.S. House of Representatives.

> Let the General Government organize the arid region, including all of the lands to be irrigated by perennial streams, into irrigation districts by hydrographic basins in such a manner that each district shall embrace all of the irrigable lands of a catchment basin and all of a catchment basin belonging to these lands, and determine the amount of water which each catchment shall afford, and then select sufficient irrigable lands for that water to serve, and declare that the waters of the catchment area belong to the designated lands and no other, and prohibit the irrigation of any other lands. In order to maintain existing rights, declare all lands irrigated at the present time to be irrigable lands. This will divide the water among the lands and prevent conflict, and rights will not grow up where they can not be maintained. Then let the people of each irrigation district organize as a body and control the waters on the declared irrigable lands in any manner which they may devise. Then declare that the pasturage and timber lands be permanently reserved for the purposes for which they are adapted, and give to the people the right to protect and use the forests and the grasses. Let the Government retain the ownership of the reservoir sites, canal sites, and headwork sites; but allow the people of each district to use them, as a body, so as to prevent speculation in such sites which would ultimately be a tax on agriculture.
>
> My theory is to organize in the United States another unit of government for specific purposes, for agriculture by irrigation, for the protection of the forests which are being destroyed by fire, and for the utilization of the pasturage which can only be utilized in large bodies; that is, to create a great body of commonwealths. In the main, these commonwealths would be like county communities in the states . . . I would have the Government declare the boundary of an irrigation district for this purpose, and then say to the people of these districts, Control these interests for yourselves. Let Congress do something more; let it say within each district, There is a body of land which is irrigable, and you can use all the water in that district on that body of land and nowhere else. Then say to

> the people, You can settle that district which is declared irrigable. You can settle that by homesteads, and that pasturage and that timber we turn over to you on this condition, that the States will agree that the people who live in any district which is divided by a State line may themselves organize their own government and use the water belonging to them as a district . . . if they allow the people to make their own laws and govern themselves in the distribution of that water, then the Government will turn over to the people of such a district the use of the timber and the use of the pasturage.[6]

In a word, the people of the districts (watersheds) would be the stewards. Powell had enormous faith in people who worked the land. He believed that they could develop a much deeper understanding of how their home watershed worked and should be accorded governance of that watershed. He himself was caught in the cultural sway of westward expansion. He was coming from an agrarian point of view but well understood that corporate industry was what fashioned "money kings," thus replacing monarchy by hereditary right.

The Civil War had been over for twenty-five years in 1890 when Powell made his presentation before the House Committee on Irrigation. The human population of the United States was still less than 75 million, and the empty West was like a honeycomb ripe for plunder.

Some of us had grandparents who were alive in 1890, myself included. I spent many childhood hours listening to my grandfathers speak of their own youth, and in retrospect, I can piece together much of the spirit of the American cultural milieu of the 1890s. American enthusiasm ran high. Rural life still predominated. American White Anglo-Saxon Protestant culture was the greatest culture yet visited upon the planet. The Union forces rightfully won the Civil War. American Indians had stood in the way of westward expansion, but they were history by the 1930s. America was abundant with natural resources. Growth was an eternal characteristic of the future of America. Mark Twain and Charles Dickens lined the bookshelves. Dinner came at midday, supper in the evening. An uncle fell through the hole in the outhouse. I met a few Civil War veterans and a Negro former slave. Hillbilly music and hymns prevailed. Old ladies quilted and cooked for church socials. Men wore straw hats, chewed stogies, and spun yarns.

When John Wesley Powell died in 1902, the planet's human population was about 1.75 billion. Today, the world population is at least 7 billion.

In those days, a doctor's house call was $2 or two chickens or less, and he arrived in a horse-drawn buggy. Coffee was a nickel. Coca-Cola was still powered by cocaine. Locomotives were coal fired. Rural populations and urban populations were about the same. America was the world's melting pot, although some flavors were more welcome than others. America was truly a comfortable place to be, fired by the enthusiasm of youthful nationalism. Them was the good old days.

What Powell envisioned was a brilliant approach to settling the West, a landscape with which he had long been intimate. He knew what watersheds were, drainage systems that started at higher elevations, be they mountain peaks or hilltops, and contained within geophysical cradles, separated by ridgelines, mountain ranges, or even high spots between arroyos, where water drained ever seaward and where biotic communities flourished or withered depending on the foibles of weather or the blessed yield of springs. He understood that easterners accustomed to limitless rivers, runs, and creeks would be flabbergasted at the barrenness of the arid West. He foresaw that in order to successfully irrigate, dams and canals would have to be constructed, irrigation ditches dug, and water rights determined where widely spaced waterholes had been defended to the death by cowboys and ranchers. And although the term *Manifest Destiny* had fallen out of favor, the seeds that it had planted a half century before had taken deep root in the cultural psyche of America. Powell stood before a congressional House committee in the late winter and spring of 1890, doing his level best to establish well-thought-out legal guidelines to prepare for the continued westward expansion that he keenly foresaw. But he was too late, and ever eager entrepreneurs were descending in waves to the extent that future legislation would result in a hodgepodge of complexity that is now both baffling and erroneous.

One of Powell's greatest fears was that interwatershed, or interbasin, transfers of water would come to prevail. Could he have possibly imagined that water would be passed from the San Juan River Watershed beneath the Continental Divide to the greater Río Grande Watershed through a twenty-three-mile-long tunnel? Or that water would be pumped from the Colorado River to coastal cities to the west and the central valleys of Arizona to the east? Powell's dream of an irrigated agrarian culture in the arid West is rapidly turning into a 3-D urban phantasmagoria wherein virtual reality seductively rearranges America's perceptions and practices.

Powell didn't expect North American Indians to survive, those peoples whose cultures had been shaped in large measure by their homelands. Thus he documented as best he could the mores and languages and artifacts of the cultures he visited, where he created enduring friendships with peoples whose minds encompassed points of view that seemed utterly alien to the sons and daughters of Europe. His own watershed perspective may well have been shaped in part by Hispano culture of northern New Mexico where irrigation was sculpted into acequia systems that expanded irrigable lands and set community political systems. Spanish land grants and watersheds were not privatized but, rather, held in common, each family garden or field allotted its fair share of water as stipulated by the elected *mayordomo*, who rostered the times when a headgate could be opened so that water could course into cultivated land. Through the centuries, countless boys of twelve years of age climbed out of warm beds in the mid of night to open headgates so that the fields of their fathers could drink, their corn, beans, squash, pumpkins, chiles, garlic, onions, and orchards absorbing life-sustaining water. Then the headgates were closed so that other headgates could be opened for their allotted times. Until after harvest, the fields were the domains of their respective family households. After harvest, anyone's cattle and sheep could wander through grazing on what was left and fertilizing the soil with their own by-products. In the arid landscape of northern New Mexico, Hispano culture survived through hard work, mutual cooperation, and faith in the benign presence of San Ysidro and other santos in the Christian pantheon that provided spiritual nurture.

In the neighboring pueblos, the Indians had long tilled their fields and devised sophisticated means of gathering sacred waters to nourish their own crops and even recharge the aquifers beneath the surface of the sacred Earth. They performed their cycles of ceremonials and danced in reverence for the local deities who presided over the weather patterns and seasonal changes, and they ever remained in awe of the mighty river whose banks their pueblos lined like gardens of humanity filled with promise of consciousness.

And in spite of westward expansion spawned east of the 100th meridian, the native cultures of the arid West continue to endure, inevitably reshaped in great measure by the dominance of the more newly arrived but still dancing to the heartbeat of the Earthmother whose song may ever be heard by those with the refined sensibilities, finely honed attention spans, and deep intuition that characterize traditional Native Americans of every cultural persuasion. This is part of their great collective contribution to human consciousness and is at least as important as science and technology. Without their wave of influence, we are indeed a world out of balance.

According to Webster, mores are "folkways that are considered conducive to the welfare of society and so, through general observance, develop the force of law, often becoming part of the formal legal code."[7] Although already complex, a far simpler system of mores dominated American culture at the dawn of the twentieth century than at century's end. Considering watersheds as commons could have become a major factor in our cultural mores had Powell been successful or had eastern Americans been more patient in the settling of the West. But by 1900, the capitalist system of unbridled private enterprise had already spread like a wildfire through drought-ridden grassland before heavy winds. The system of ethics that was embedded in Powell's watershed master plan would not be redefined for another half century when the great conservationist Aldo Leopold included his profound essay "The Land Ethic" in *A Sand County Almanac*. Therein Leopold wrote, "Quit thinking about decent land-use as solely an economic problem. Examine each question in terms of what is ethically and esthetically right, as well as what is economically expedient. A thing is right when it tends to preserve the integrity, stability, and beauty of the biotic community. It is wrong when it tends otherwise."[8] Imagine that Leopold used the word *watershed* rather than *question* so that it would have read, "Examine each *watershed* in terms of what is ethically and esthetically right, as well as what is economically expedient." In other words, the watershed is itself a commons for the biotic community that it cradles and sustains.

What would it take to reintegrate the acceptance of watersheds as commons into the mores of global monoculture? Remember the popular apothegm, "Think globally. Act locally." Understanding watersheds as commons is a mighty leap in a culture where water law in the American West is cast in slickrock, where certain legislation is revealed to be at loggerheads with natural law, where selected cultural mores were molded into law as an economic expedient rather than following the path to refinement of ethical standards. Imagine integrating the best thoughts and practices of John Wesley Powell and Aldo Leopold into a "Leopowellian" masterpiece of bioregional endeavor.

This could indeed work in many watersheds that appear in Powell's great map, although the Colorado River greater watershed is now subject to an overwhelming array of factors and serves a human population that has now exceeded the carrying capacity of the Lower Basin of the Colorado River. Consider that Powell died when Leopold was fifteen years old and the population of California was but one-thirtieth its current size. Consider that the human population served in the Lower Basin of the Colorado River is now greater than 25 million with more on the way on a daily basis. Consider that the plumbing of this incredibly complex miasma of cities and agricultural districts occurs by interbasin transfers of water that Powell had hoped to thwart so that each of the watersheds of the West could retain its respective integrity. Consider that instead of being governed from within by people familiar with the workings of their respective watersheds, the Lower Basin of the Colorado River has as its mayordomo the personage of the secretary of the interior, the moderator of a riverine empire so populated by humans that the rest of the biotic community has fallen out of sight, a river system so overallocated that scarcely a drop reaches its natural destination—the Sea of Cortez, itself a magnificent ecosystem that nourishes countless fellow species and delights the souls of all who have played in its waters, camped on its islands, and hung out with and recorded geomythic mapping songs of its indigenees.

WHEN ONE LOOKS AT THE MAP OF THE ARID LANDS OF THE WEST provided by Powell for the U.S. Geological Survey's *Eleventh Annual Report*, one see about 150 different watersheds represented in different shades of color and subtly different hues. In the American Southwest, which for our purposes includes New Mexico, Arizona, southern Colorado, southern Utah, southern Nevada, and southern California, the major watersheds include the Arkansas-Canadian, the Río Pecos, the Río Grande, the Gila, the San Juan, the Colorado, and the Great Basin (closed).

The Arkansas and Canadian rivers' headwaters are in Colorado, and their land base includes parts of Colorado, New Mexico, Kansas, Texas, and Oklahoma, where they conjoin to form the main stem of the Arkansas River that flows into the Mississippi River in Arkansas. They gain momentum as they flow eastward, but they are modest in their flows closer to their respective sources, where they pass through lightly populated country until they approach the 100th meridian.

Southwest of the Arkansas-Canadian River Watershed is the Río Pecos Watershed that begins in northern New Mexico and finally drains into the Río Grande near Langtry, Texas. In New Mexico, the Río Pecos passes through Pecos, Santa Rosa, Fort Sumner, Roswell, Artesia, and Carlsbad before crossing into Texas and passing through the communities of Pecos and Langtry, where during the nineteenth century, Judge Roy Bean proclaimed himself "the law west of the Pecos" and where he held trials in his saloon.

Farther west is the watershed of the Río Grande, whose headwaters spring forth in the San Juan Mountains, head south through the San Luís Valley of southern Colorado, and then bisect New Mexico east from west before entering Texas at El Paso where the river becomes the international border between Texas and Mexico until it empties into the Gulf of Mexico. The Río Grande is 1,896 miles long and drains an area of 182,200 square miles. The northern Río Grande drainage area extends from its headwaters near Creede, Colorado, south to Fort Quitman, Texas.

The western boundary of the Río Grande Watershed is the Continental Divide, east of which all waters eventually drain into the Atlantic and west of which they either drain into closed basins such as the Great Basin Desert of Nevada and Utah or the Pacific Ocean and Sea of Cortez.

The Colorado River Watershed includes the Green, San Juan, Animas, Escalante, Virgin, Paria, and Bill Williams rivers that drain 244,000 square miles in parts of Colorado, Wyoming, Utah, New Mexico, Arizona, Nevada, California, and 2,000 square miles of Mexico.

The headwaters of the Gila River Watershed occur on the western slopes of the Continental Divide in southwestern New Mexico. The river flows east to west through southern Arizona until it drains into the Colorado River at Yuma, Arizona. With the signing of the Treaty of Guadalupe Hidalgo in 1848, the 650-mile-long Gila River marked the boundary between Mexico and the United States. After the Gadsden Purchase of 1853, a wide swath of landscape south of the Gila became the last land acquisition within the area of the lower forty-eight states.

These five watersheds span the American Southwest. All of the waters that pass through the rivers and streams in these five watersheds amount to but a small fraction of the yield of the Columbia River in the Northwest. The Southwest is the driest part of America and remains the emptiest in human terms, although its urban centers are among the fastest growing in the United States. At the same time, produce grown and irrigated in the Imperial Valley and other areas in the Lower Basin of the Colorado River feeds millions of Americans and is a major source of revenue. At least 80 percent of New Mexico's water is slated for irrigation. It is obvious to all that agricultural and urban areas now vie for the modest waters of the

American Southwest. Water laws that were passed in the early twentieth century cannot accommodate the complexity of present and future demands.

IN 1922, REPRESENTATIVES OF SEVEN WESTERN STATES GATHERED AT Bishop's Lodge just north of Santa Fe to address apportionment of the waters of the Colorado River Watershed. The meeting was presided over by Herbert Hoover, then secretary of commerce. In previous years, the Colorado River had been running high, and subsequently folks erroneously thought that it was safe to assume that the river could be counted on to deliver 17 to 18 million acre-feet per year. Thus representatives from the seven states generally agreed that the watershed would be divided into two basins, the Upper and Lower Basins, the dividing line occurring at Lee's Ferry between Glen Canyon and Marble Canyon in Arizona. Each basin was to be allocated 7.5 million acre-feet annually. The Upper Basin states include Wyoming, Colorado, Utah, and New Mexico (and a slice of Arizona); the Lower Basin states include Arizona, California, and Nevada. The Upper Basin states were agreeable to apportioning the waters between themselves. The Lower Basin was considerably more complicated because California wanted to claim the lion's share of the water even though it contributed virtually no tributarial flow. Arizona took great umbrage at this, and Nevada was all but the odd state out. It was determined that California would receive 4.6 million acre-feet per year, Arizona 2.3 million acre-feet, and Nevada but 300,000 acre-feet per year. In 1922, there was very little happening around Las Vegas, although the Las Vegas of today is immense and looking for water.

Arizona did not agree to the Colorado River Compact, thus it took four decades and a U.S. Supreme Court decision to determine that Arizona would be allocated 2.8 million acre-feet from the Colorado River and all of its instate tributarial waters. Arizona remains junior to California so that if the river runs low, California gets its water first.

Neither Indians nor Mexicans were invited to Bishop's Lodge, and it wasn't until 1944 that the United States entered into a treaty with Mexico guaranteeing Mexico 1.5 million acre-feet a year. Between the Upper and Lower Basins and Mexico, that brought the total acre-feet to be apportioned up to 16.5 million. In the meantime, it was revealed that the average rate of flow was much closer to between 13.5 and 15 million acre-feet, not the 17 million that was previously thought.

California initiated a massive water storage project that resulted in construction of the Hoover (originally Boulder) Dam that was dedicated by President Franklin D. Roosevelt in September 1935. Lake Mead has the capacity to store 28.5 million acre-feet and is located about thirty miles downstream from Las Vegas, Nevada. However, since 2000, it has been dropping, and by the time of this writing in 2010, it had reached a record low and holds less than 40 percent of its capacity.

Downstream from Hoover Dam is the Imperial Dam that diverts water from the Colorado River into the Imperial Valley of southern California and to several cities. Parker Dam is situated between the Hoover and Imperial dams and impounds Colorado River water into Lake Havasu, which is the source of the Colorado River Aqueduct that provides much of southern California's drinking water.

As the twentieth century unfolded, it was apparent that the Colorado River might have 25-million-acre-foot years and 6-million-acre-foot years. By virtue of the 1922 Colorado River Compact, the Upper Basin had to guarantee 7.5 million acre-feet of Colorado River water each year, or 75 million acre-feet over any ten-year period. Thus the Colorado River Storage Project (CRSP) was conceived in 1956 to develop storage and irrigation facilities in the Upper Basin that resulted in construction of several dams, including the most contested dam in America, the Glen Canyon Dam, that is situated just eighteen miles upstream from Lee's Ferry, the dividing line between the Upper and Lower Basins of the Colorado River.

The Glen Canyon Dam backs water up into Glen Canyon, resulting in Lake Powell, named after John Wesley Powell. One can but wonder what Powell would have thought of this enormous reservoir with a capacity of over 24 million acre-feet. We know what author Edward Abbey thought and can read all about it in his classic novel, *The Monkey Wrench Gang*. In theory, Lake Powell guarantees that the Upper Basin can indeed release 8.23 million acre-feet a year through the Grand Canyon into the Lower Basin reservoir, Lake Mead. That amount includes both Lower Basin water and half of America's commitment to Mexico by virtue of the 1944 treaty.

The Glen Canyon Dam was dedicated in September 1966 by Lady Bird Johnson before a host of leading politicians. Thus, one of the most beautiful canyons in the world drowned beneath Lake Powell.

As author William deBuys points out, the noted historian Samuel P. Hayes well characterized the milieu of the turn of the twentieth century and subsequent decades as a time when America was motivated by "the gospel of efficiency."[9] Hayes spoke authoritatively about the effect of the Industrial Revolution on American culture. Indeed, anthropocentrism made little room in cultural consciousness for concern for other biota or the balance of natural forces. When the Colorado River Compact of 1922 was put into effect, no thought was given to the river for its own sake. It was there for human use, period. Concern for habitat was paltry compared with the drive for endless economic growth. The Colorado and other rivers were there to be put to best use, as agriculture was then regarded. A culture religiously motivated by the right to dominate as defined in the biblical book of Genesis was appalled by the voice of Charles Darwin, who possessed one of the greatest minds of the nineteenth century.

For years, Arizonans had been craving their own enormous public works project, one they named the Central Arizona Project (CAP). In the early 1900s, the Salt River Project was initiated and resulted in construction of the Roosevelt Dam at the confluence of Tonto Creek and the Salt River. Roosevelt Lake became the reservoir that would provide water for irrigation to farmers around the Phoenix area. Ultimately, the Roosevelt Dam was to also generate hydroelectric power to provide electricity for the region. At the time of its construction, Roosevelt Dam was the tallest concrete dam in the world, standing 280 feet high. It was completed in 1911. The Salt River Project was so successful that Arizonans envisioned an even greater public service project, the CAP.

If more water could be sprinkled on the sands of the Sonoran Desert for agriculture, Colorado River water could be put to what was still deemed best use. The concept involved pumping waters east from Lake Havasu (on the other side of the lake from the pumps that water cities in southern California) into an aqueduct that ultimately stretched for 336 miles, all the way to Tucson. Originally, electricity to run the pumps was to have come from two "cash register dams" to be located at either end of the Grand Canyon. Martin Litton and David Brower of the Sierra Club were instrumental in stopping that project before it began. However, electricity had to come from somewhere.

The largest coal deposit in Arizona lies buried in the heart of Black Mesa, a land formation sacred to both Hopi and Navajo Indians. The new plan was to strip-mine coal from Black Mesa, transport it via a specially constructed railway to the shores of Lake Powell where the Navajo Generating Station was to be constructed, thence to fire electricity through power lines supported by towering metal kachinas from Lake Powell to Lake Havasu to power the pumps to pump the water into the aqueduct to the central valleys of Arizona. As an adjunct, additional coal was to be slurried with water pumped from the Pleistocene aquifer beneath Black Mesa from the mine through a 273-mile-long slurry line to the already existing Mohave Generating Station near Laughlin, Nevada, to help light up Las Vegas.

While the operation was to provide jobs for Indians, the blow to Hopi traditional culture, and to many traditional Navajos, was to prove nearly lethal. Black Mesa has now been strip-mined by the Peabody Coal Company of East St. Louis for nearly four decades in spite of attempts to thwart the debacle by environmentalists who were asked to help by Hopi traditional elders. The rape of Black Mesa stands as perhaps the greatest model of complete environmental and cultural catastrophe to have ever been visited upon the American Southwest by "captains" of industry and their men in government. Indeed, it is here that the seeds of Manifest Destiny achieved a most complex incarnation.

Ironically, the $5 billion price tag for the Central Arizona Project rendered the water far too expensive for farmers. Thus many farmers have sold their farms to land developers and other modern-day carpetbaggers, and Phoenix and Tucson continue to spread across the fragile Sonoran Desert. Does this signal the shift from agriculture to urban growth as the new interpretation of "best use" of water in the American Southwest?

THE COLORADO RIVER COMPACT OF 1922 DETERMINED THE IMMEDIATE fate of the Colorado River, the lifeline of the Southwest, with what is now known as "the law of the river." Procedurally, it set the tone and guidelines for other compacts, including the Río Grande Compact that defined the river rights of Colorado, New Mexico, and Texas.

Today, most of a century later, many realize that the river compacts of the early twentieth century have led to enormous new problems based on human population growth and early lack of ecological consciousness, both problems that continue to persist. Native American water rights have yet to be decided. The notion of "use it or lose it" continues to color everyone's thinking. Prior rights seem cast in stone. Groundwater is being pumped out of aquifers faster than it can be recharged.

For example, the city of Albuquerque was thought to be situated over an aquifer the size of Lake Michigan. Then in 1982, that was revealed not to be the case; rather, the subterranean water supply was being depleted faster than it could recharge. The city would have to get water from the Río Grande that was already allocated in favor of agriculture, thus pitting farmers against the city. A deal was struck earlier on to create an interbasin transfer of water from the San Juan River Watershed (which is part of the greater Colorado River Watershed) that would be pumped beneath the Continental Divide into Heron Lake on the Río Chama to thereafter flow into the Río Grande. It could thence be stored in a series of reservoirs culminating in Cochiti Lake, the river stoppered behind an immense earth-filled dam about thirty miles upstream from Albuquerque, and thus be made available for urban use.

This is precisely what Powell had hoped to avoid, recognizing that water use should be restricted to what was available within the watershed and that interbasin transfers were wrong on all counts. Indeed, watersheds should serve no more than their own denizens. Interbasin transfers result in falsely defined watershed carrying capacities and jeopardize those who grow to rely on imported water. As attorney and author Em Hall points out, Albuquerque is now in direct competition with Los Angeles, Phoenix, Tucson, Las Vegas, and other cities for Colorado River water, "and who's going to win in that kind of competition?"[10]

We have expanded beyond the parameters of human well-being with little regard for the well-being of our fellow biota here in the American Southwest. We have foregone mindfulness of the nature of home habitat. While environmental consciousness has grown in magnitude over the last forty or so years, it has not kept pace with economic growth, population growth, industrial growth, extraction of nonrenewable resources, or an increase in pollution. It's as though we are culturally polarized to the point where we cannot thwart our growth because we cannot agree to see with the clarity necessary to turn the juggernaut of our own creation. And we have centralized political power to the extent that our political bodies are ripe to be swayed by those in economic power. At this point, they are one and the same, an aggressive entity that holds sway over cultural consciousness and otherwise controls everything, unchecked, even to the point of denying the undeniable—namely that we are soon to be brought to our knees because of seven generations of massive overindulgence at extreme expense to seven generations hence.

The problem is further exacerbated in that we have allowed much of our cultural consciousness to be subsumed by "virtual reality" through addiction to the "digital mind." This is actually a fascinating phenomenon—to watch evolved minds become and remain dominated by BlackBerries and other fruits of the digital generation, now approaching a quarter century since coming of age. What's tragic is that so little of the digital mind is relevant, so much of it electronic persiflage. To think that we talk to a

BlackBerry rather than a grandfather juniper tree or a range of buffalo grass. We listen through headphones to engineered sound rather than to the call of the coyote, the song of the meadowlark, or the cry of a hawk. Most of us separate ourselves from that which sustains us—we rarely swim in the flow of Nature.

So what's next in this stream of consciousness?

A lot of it has to do with common sense.

Fifty years ago, I had a friend, Rick Mallory, who foresaw that at some point, tough times are a comin'. Rick's answer was to buy a piece of countryside complete with hot spring and build a modest house adjacent to it so that the spring could provide heat as well as water. Then grow a sufficiently large garden and hunt venison, elk, antelope for meat. Jerk it, smoke it, and eat it.

I had another friend, Ed Abbey, who wanted to buy enough rangeland to put a herd of bison on it and invite a few friends to set up a community—each household separated from the rest to afford privacy—enough of a community so that when the time was right, a great bonfire would be set alight, a haunch of buffalo roasted, homemade brew provided in abundance, turnips cooked, bread baked, and a great fiesta held, complete with homegrown music, poetry, and nights filled with lovemaking beneath a dark firmament filled with star-bright constellations reflecting mythic moments of yore.

Brother Abbey was indeed a man of refined taste.

My own youthful vision of perfect survival in days to come was to buy an inholding surrounded by national forest near a river, the forest surrounded by Indian reservations, the landscape filled with deer, elk, and the occasional bear and mountain lion, complete with ample firewood, a good well, a garden, and an orchard—the only concession to luxury being a library of fine books that never staled and plenty of time for walking and meditation and running wild rivers.

Some of us had long been enthralled with life as a hunter-gatherer and even practiced it in days of youth. The great thinker Paul Shepard justified this sentiment in his compelling book, *Coming Home to the Pleistocene.*

But these fantasies could only occur in a far less populated world. Over the last half century, I've been honored to befriend many Indians—Native Americans, people indigenous to their homelands, there rooted by dozens, scores of generations of forebears.

From as far north as the Nez Perce people of the Clearwater Watershed, to the Mayans of Chiapas in southern Mexico, from each of the peoples in between who still retain their traditional values and points of view whom I've visited, recorded, even lived with for months of my life, I've received clear understanding that this soil we tread is sacred soil, sustaining soil to be revered. The scent of juniper smoke in the Southwest aligns our sensitivities in harmony with homeland. Their local deities are profoundly in place and celebrated in ceremonials throughout every tribal domain. Their songs bespeak mythic history that retains its presence in the moment. States of mind are available wherein one becomes aligned with other ways of plumbing the mystery of existence. Their homelands, reserved for them by virtue of treaties that *must not be broken*, are gardens of survival potential. They are the inheritors of a greater landscape that is now tired out by centuries of overuse by newcomers who still don't understand what it takes to think like a watershed. Our children share this heritage. It is now up to our children of all cultures to save us from ourselves.

MUCH THAT IS WONDERFUL HAS BEEN GIVEN TO THE WORLD BY WESTERN culture: classical music since the time of the Renaissance, the paintings and sculpture of master artists, extraordinary literature rendered in both poetry and prose, philosophical speculations that challenge conflicting absolutes, Western medicine, science in its myriad areas of discipline and application, technology in its seemingly infinite ability to remanifest itself in response to perceived human needs, digital media that present a most seductive virtual reality and a new means of participating in evolving cultures of practice.

However, Western culture has been costly to the earthly habitat that spawned us humans into the keystone species. And dominant global human culture has now absorbed much of the offerings of science and technology of the West into an extraordinary multicultural complex that lies far beyond the ken of any single human mind or institution to encompass. In my mind's eye I envision this planet, itself a living organism spinning along its ancient trail through space around the sun in a multibillion-year-long steady state that nurtures life and its potential. I perceive much of burgeoning humanity engaged in what seems a frenetic expenditure of energy that serves no apparent purpose other than pursuit of industry within territorially defined regions marked by ever increasing wars supported by a faltering environment. I also perceive patches of survival potential in habitats that struggle to preserve their own integrity wherein clusters of humans still hold the planet in awe and reverence for its ultimate gifts of life and consciousness. Some of these clusters struggle to prevail in the American Southwest, where Indians of different cultural persuasions, who are descended through countless generations within cultures that evolve as their respective habitats evolve, sing their wisdom to the rhythm of heartbeat in homeland.

Myths and legends are not simply superstition couched in creation stories and songs celebrating earth forms and fellow creatures. They are expressions of intuitive understanding of homeland, of vast kinship, of immersion within the flow of Nature. They are lore-filled accounts of generations of observation absorbed into cultural encyclopedias of oral history arranged not alphabetically but by subject as perceived within the present. They are reenactments of moments of cultural truth. They are braids of metaphor entwined within an umbilicus that adheres culture to homeland. They evoke sensibilities that plumb the entirety of human understanding in a way that intellect alone cannot. They provide clear insight into perceiving watersheds as commons—because that's what watersheds are, habitats common to all species and geophysical forms therein. It does not require an advanced degree in physics to understand that. Rather, it takes an entire array of refined sensibilities to discern the persistence of the sacred and to guard against secularization of homeland. In no way does this invalidate the practice of science. Rather, it provides a much wider and deeper purview from which to intellectually seek meaning in existence.

THE LATE, GREAT CULTURAL ANTHROPOLOGIST EDWARD T. "NED" HALL burst through the front door of our home in Santa Fe shouting, "Jack! We need to create an ethnic global jukebox!" This was in the early 1980s, well before the Internet made such a concept a potential reality. Ned was on the mark. In a recorded interview, Ned had this to say:

> The land and the community are associated with each other. And the reason they're associated and linked, and the reason that people get their feeling of community from the land, is that they all share in the land. Ethnicity is looked upon normally as a liability, because people want to make everyone else like themselves. And this is something we're going to have to learn to overcome, because ethnicity is one of the greatest resources, if not *the* greatest resource, that we have in the world today. What we have here are stored solutions to common human problems, and no one solution is ever going to work over a long period of time, so we need multiple solutions for these problems. So ethnicity is like

money in the bank, but in a world bank. Culture is an extension of the genetic code. In other words, we are part of Nature ourselves. And one of the rules of Nature is that in order to have a stable environment, you have to have one that is extraordinarily rich and diverse. If you get it too refined, it becomes more vulnerable. So we need diversity in order to have insurance for the future. Again, you need multiple solutions to common problems. The evolution of the species really depends on not developing our technology but developing our spirits or our souls. The fact is that Nature is so extraordinarily complex that you can look at it from multiple dimensions and come up with very different answers, and each one of them will be true. And we need all of those truths.[11]

Implicit in Ned's wisdom is that the totally rational mind is blind to the scope of perception of which we as humans are capable. Yet without rationality we are bereft of the capacity to extrapolate our way through the complexity of Nature. A great truth is that rationality and applied technology alone will not see us through the coming decades.

Yet without them, we will not survive.

The mystery of existence, life, and consciousness is to me the most compelling and endless realm of pursuit. There are many trails to the heart of truth, many methods of pursuit. As our species evolves, our perceptions of the mystery grow more complex until it becomes obvious that there is no fixed way through which the mystery may be seen and understood in its entirety. At best, we catch glimpses through tools of science or by meditation, ceremony, prayer, observation, or even surrender to the flow of Nature, to name but some "doors to perception." Belief in any one system to the exclusion of others is to blindside ourselves. Conflicting absolutes, such as pitting scientific method against organized religion, create a mire that only serves to impede clarity. Yet it is difficult to metaphorically reside within a crystal of many windows and perceive the myriad views simultaneously.

In an earlier work, I wrote the following:

Living a creative self-directed life is like running a wild river; you don't deny the current its due, but you work your own way through the rapids, camp where you will, explore side canyons that intrigue you, and relish the danger, heeding no higher authority than the truth.[12]

Thinking like a watershed comes naturally to me; I could easily be regarded as being devoid of focus. In my own haphazard fashion, I have pursued exposure to as many points of view as I could during this adventure of life. I haven't restricted my adventure to the American Southwest, but it is definitely my region of deep preference. I'm at home anywhere in the rural Southwest and can happily adapt to any place west of the 100th meridian in the contiguous forty-eight states of the United States, and even the desert country of northern Mexico. I am a desert rat fascinated by waterways, including dry arroyo bottoms that afford temporary storylines scribed in the sands. I love to run rivers through the high-desert country or hike canyons, camping in caves, sharing space with Tarahumaras, looking into a grandfather juniper tree momentarily populated by a wave of mountain bluebirds passing through, stopping for a quick snack. I love to listen to, even record, the sounds of Nature, including points of view and songs of Ute, Navajo, Puebloan, Hopi, Zuni, O'odham, Apache, Yaqui, Seri, Mayan, Huichol, Basque, Hispano, Buckaroo, and other ethnic communities who have achieved indigeneity to homelands. To listen to the dawn chorus of birdsong near a spring in the Sonoran, Great Basin, Mohave, or Chihuahuan deserts is to hear an expression of consciousness that is not alien but deeply kindred. To listen closely to the hiss of a rattling diamondback gives great meaning to biosemiotics, Nature's system of signage. To wander through the geophysical cradles of watersheds, listening to the drone of insect wings, is to experience a biophony of sound more sacred to me than a baroque oratorio. To hear the cry of a red-tailed hawk or the mellifluous song of a canyon wren is to experience a form of perfection rendered by Nature.

Each watershed is populated by a biotic community that strives for collective balance with homeland. Human presence is part of almost every biotic community, and as the present keystone species on our planet, we humans must pay close heed to engaging in "communities of practice" that don't violate natural balance or endanger it. Nature has a mind of its own that human intuition can sense only if we guard against allowing that intuition to atrophy in our quest to fulfill other human urges or otherwise succumb to self-absorption.

What follows is a presentation of different human points of view from within several watersheds of the American Southwest. It is a presentation of gifted minds from different cultures that needn't articulate watersheds as commons for respective biotic communities, because that understanding is tacitly shared.

This anthology is offered with humility in the hope that it may in some small way contribute clarity to the broader spectrum of human perception.

Notes

1. Garrett Hardin, "The Tragedy of the Commons," *Science* 162, no. 3859 (December 23, 1968): 1243–48.
2. Thomas Malthus, *An Essay on the Principle of Population* (London: J. Johnson, 1798).
3. Donella H. Meadows, Dennis L. Meadows, Jorgen Randers, and William W. Behrens III, *Limits to Growth* (New York: Universe Books, 1972).
4. Jack Loeffler, *Survival Along the Continental Divide: An Anthology of Interviews* (Albuquerque: University of New Mexico Press, 2008), 25.
5. John O' Sullivan, "Annexation," *United States Magazine and Democratic Review* 17, no. 1 (July–August, 1845): 5–10.
6. U.S. Geological Survey, *Eleventh Annual Report* (1888–1890), 256–57.
7. *Webster's New Universal Unabridged Dictionary*, s.v. "mores."
8. Aldo Leopold, "The Land Ethic," in *A Sand County Almanac* (New York: Oxford University Press, 1949), 224–25.
9. William deBuys interview by Jack Loeffler, in "Getting to Know the River," program 1 of *Moving Waters: The Colorado River and the West*, a six-part radio series, 2002, produced by Jack Loeffler for the Arizona Humanities Council.
10. Em Hall interview by Jack Loeffler, in "Consciousness, Conscience, and the Commons," program 14 of *Watersheds as Commons*, a fourteen-part radio series, 2011, produced by Jack Loeffler and Celestia Loeffler.
11. Edward T. Hall interview by Jack Loeffler, in "Manifest Destiny: The Coming of Anglos to the American West," program 2 of *Watersheds as Commons*, a fourteen-part radio series, 2011, produced by Jack Loeffler and Celestia Loeffler.
12. Jack Loeffler, *Adventures with Ed: A Portrait of Abbey* (Albuquerque: University of New Mexico Press, 2002), 3.

Rina Swentzell, photo by Jack Loeffler

Rina Swentzell

RINA SWENTZELL WAS BORN INTO THE CELEBRATED NARANJO FAMILY IN the Tewa-speaking Santa Clara Pueblo. She earned her BA in education and her MA in architecture from New Mexico Highlands University. She earned her PhD in American studies in 1982 from the University of New Mexico. Swentzell writes and lectures on the philosophical and cultural basis of the Pueblo world and its educational, artistic, and architectural expressions. Her writing appears in magazines, scholarly journals, and edited collections, and she appears in video presentations for television and museums commenting upon Puebloan cultural values. She has been a consultant to a number of museums, including Santa Fe's Institute of American Indian Arts and the Smithsonian, and was a visiting lecturer at both Yale and Oxford in 1996. She has contributed to Lore of the Land as a board member and scholar since 2007. She is the author of *Children of Clay: A Family of Pueblo Potters* and *Pottery in Santa Clara: A Photographic History of Pottery in Our Community*. She is coauthor of several other books. In the pages that follow, Rina Swentzell reflects on the Puebloan way of thinking about and perceiving home watersheds in the arid landscape of northern New Mexico, where her culture has gradually evolved in response to the needs of the homeland for countless generations.

Pueblo Watersheds

Places, Cycles, and Life

IT WAS A QUIET, EARLY SUMMER DAY, AND WE, THE SANTA CLARA PUEBLO people, were solemn and thoughtful. There had been no rain for a couple of months. Rivers and streams were below their normal flows. Plants were shriveling and the ground was packed and crusty. Any loose dirt had long ago been blown in half-pyramid forms into the corners of the plaza. It was time to ask for consideration from forces beyond our human selves. It was time to ask the cloud people for help. It was time to ask for their love—to ask for rain.

As was already happening in many of the other Pueblo communities throughout northern New Mexico, we stood surrounding the double row of dancers with their headdresses moving against a clear blue sky. The singing was tender and strong with feeling. There were no visitors; only community people stood watching and joining their energies with the song and movement of the dancers. For most of the day, the people were quietly focused. Then, during the third coming-out of the dancers, a small cloud, seemingly out of nowhere, formed over us. Shortly, tears joined the gentle drops of rain that fell on our faces. The cycle of life had once again embraced us.

Definition of Watershed by Pueblo People

There is the understanding in the Pueblo world that a watershed is a whole cycle of water movement within our natural world that includes the skies, the clouds, the mountains, the hills, the surface waterways, and the groundwater beneath the surface—as well as humans, plants, and other creatures. This was the preeminent thought offered during the sixteen interviews for the Lore of the Land watershed project that my sister, Tessie Naranjo, and I conducted with Pueblo people over a two-year period. Although it is well acknowledged by Pueblo people that our surface waters originate in the far and surrounding mountains, the people we interviewed insisted on looking to the skies, to the clouds, as the beginning of any watershed cycle. A Tesuque Pueblo woman talked about the cycle beginning in the skies. Rain, she says, "comes from the sky, goes into the Earth, underground, comes back up into lakes and rivers, and then evaporates back into the clouds." This expands the general definition in Western thought that a watershed is an area of land that drains into a lake or river. Or that a watershed is a ridge of high land dividing two areas that are drained by different river systems.

This two-dimensional view of a watershed within general Western thinking is expanded into three dimensionality by the Pueblo interviewees when they speak in terms of a cycle of movement in which water, people, and life are included. This thinking is closer to John Wesley Powell's statement that "a watershed is that area of land, a bounded hydrologic system, within which all living things are inextricably linked by their common water course and where, as humans settled, simple logic demanded that they become part of a community."

In the Pueblo world—and in John Wesley Powell's thinking—*community* and *watershed* are synonymous. A watershed is an interwoven web of life energies from clouds to rivers to streams, springs—and tears. The analogy goes further when a woman from Jemez says, "We [humans] are like clouds. Our bodies are 99 percent water." So our human songs, dances, and thoughts can communicate with the clouds because, as a San Felipe Pueblo woman said, "Water can talk with water." Many of the Pueblo interviewees viewed water as the very lifeblood of both the people and the natural community.

In this way of thinking, "The water in a watershed and the blood flowing through the human body are doing the same thing," said the woman from San Felipe Pueblo. They sustain and nourish life. They help create organic systems where connections and communication between parts and other wholes can happen. Any watershed and any human body are both a whole and a part of another whole at the same time. And they do define where we live. Both are home, each is where we live.

As each human body is unique, so is each watershed. The uniqueness of any watershed is defined by the surrounding landforms such as mountains, hills, and valleys. "There are no geopolitical boundaries," noted a Tesuque Pueblo man. And, in Pueblo thinking, a great emphasis is placed on mountains as definers of world space, or places where we live, for example, watersheds. The Okhuwa, or cloud people, arise from the mountains and move through the adjoining hills and valleys. That movement is like the breath of the universe. A Tewa prayer begins, "Within and around the earth, within and around the hills, within and around the mountains, the breath returns to you." This breathing cycle is related to water in a statement by a San Juan woman who simply says, "Water is part of the breathing cycle." But, as there is one universal breath, there are also breaths peculiar to any place. In some places the breath, or wind, moves gentler than in other places because of the whole nature or physical formation of that region. Water, similarly, moves down mountains and through hills differently than it does in valleys. A man from Santa Clara Pueblo says, "There is fast water and slow water." He goes on to say that the Río Grande has slow water, and the mountain stream that flows through Santa Clara Canyon is fast water, and they must be treated and talked with differently. Moreover, any water brings with it the soil, the ground, that it moves through, making the quality of water in any watershed unique. Then, as we drink that water, our lifeblood, we are again and always one with and part of that place, part of that watershed, part of that unique community.

Meaningfulness of Water and Place

"The prayers that we [Pueblo people] say show the spiritual significance of water, and just lately it's been proven through scientific mythology that when you talk nicely to water, water looks pretty," said a man from Tesuque Pueblo. A couple from Jemez Pueblo talked with enthusiasm about water bringing beauty because the whole cycle of rain coming from the clouds and touching rocks, plants, and creatures expresses our common inherent beauty. "Water brings beauty to the land, the grass, and [hence, it also] brings happiness. It brings beauty because it brings out all the colors in the earth and the trees." This thinking implies that water reminds us (humans) that we are also beautiful because we are one with both water and place.

For Pueblo people, water and place are also inseparable. Where water emerges from the ground is of great significance. Shrine places are often located there. They are symbolic openings for the emergence of all life because sometimes there is no visible water source. These watery places are where we connect with the underworld, that place that we came from and to which we return when we die. It is where the people "go back to the lake, back to the place of origin, back to the watery place. In that place, we breathed water and not air," stated a young man from Santa Clara.

A Tesuque Pueblo woman talked about springs as "being the opening into the womb" of Mother Earth.

There are shrines at these watery places of origin. Generally, there are four major shrines located in the far mountains surrounding any village. Then, there are the shrines located in the close hills forming another of the concentric circles embracing any Pueblo village. The next ring of shrines is around the outer limits of the physical village. And the center point is in the Pueblo plaza. "The center is our community. That first ring is our homes. The second ring is the foothills, and the third and fourth rings, all the way to the top of the mountains," said the man from Tesuque Pueblo. All these rings have shrines where connection can be made with the natural energies of that place, including water. And the clouds and rain from the sky, in combination with these horizontal concentric circles around place, define a world sphere, a three-dimensional watershed, within which a whole community of plants and animals, including humans, can exist in an interdependent and healthy way.

Moreover, the Tesuque Pueblo man continued, "Our villages are often built next to springs. These springs are connected to surface waterways, and these waterways are connected to the lakes and mountains, which in today's terms are called watersheds." Springs and lakes help define particular community or watershed areas in Pueblo thinking. They connect to the water that flows underground. Moreover, the Okhuwa, or cloud people, come from those watery places in the mountains. "It is from these places that the people-gone-on-before sent clouds and rain. They [the cloud people] are the manifestation of thoughts and desires of those gone on before," continued the young Santa Clara person. The cycle of life, then, goes from underground water sources, to surface water places, and into the clouds in the sky. Healing for humans happens when we are part of that cycle. Water is collected from all those water places for healing ceremonies. Rivers, as well, are places of cleansing and healing because the water is moving and the energy is fresh. Spring, lake, and river waters are brought into the living human community as connective blessings from the larger region, the larger water place where our ancestors dwell.

In that sense, all of the interviewees emphasized the sacredness of water and its spiritual significance. A number of them simply said, "Water is sacred." The Jemez woman reiterated this by saying that "everything comes from and goes back to water. There is no life, as we know it, without water." It is the source of life. Another woman talked about our parents being water and earth (clay). The intertwining of life elements is a critical philosophical assumption in the Pueblo world.

Therefore, Pueblo prayers and songs, as attested by the Pueblo voices, are conversations with the clouds. A man from Jemez Pueblo talked about how he makes songs. "I talk to the beautiful clouds who are forming in the northeast from where the rain is going to come. And then the next part is to say that the clouds are here and it is raining." A young man from Santa Clara Pueblo translates a song, "Far away to the north, Okhuwa boys rise up with fog and clouds and together they are coming here. Far away, Okhuwa girls rise up with lightning, thunder, rainwater, and rain splashing. All together they are coming here." It is always a telling, a story about how life continues to happen.

And, of course, song brings the dance, which is the touching of the earth with the feet and the reaching of the head into the sky realm with the mountain and cloud *tablitas*. The human, in the dance place, is the connector of Earth and sky. And it is all about water. Pueblo dances and songs are about collecting water from the skies.

Ways of Using Water in the Traditional Pueblo World

One of the daily activities in the traditional Pueblo world was to manage the surface water that flows through the mountains, hills, and valleys. A man from Tesuque Pueblo said, "Our ancestors manipulated the landscape by placing rock [grid] structures for the harvesting of precipitation from the mountains all the way down to the bottomlands. These structures manipulated the landscape so that the land was used as a sponge, allowing moisture to reach the deep-rooted vegetation and also recharging the aquifers. Our ancestors managed the surface water into groundwater." He continued talking about how this happened during the course of a day in the pueblo. "The war chief would tell the men from the west side of the village, 'All you men from the west side, we're going to go up into the foothills and we're going to build these structures.' Other men were left to work in the fields—or go hunting."

In and around the old Pueblo prehistoric sites are evidences of this kind of collection and management of water for drinking and irrigating. Tsankawi, in the Jemez Mountains of northern New Mexico, for instance, was mentioned as a possible water institute by the interviewee from Tesuque Pueblo. Tsankawi was contemporaneous with Bandelier, Puje, and the other Pajarito sites. The site is located on a tufa rock mesa on the west side of the Río Grande. The entire mesa and surrounding area are channeled with hand-chiseled grooves to move water into small and large collection places. Huge boulders are placed to create water channels to direct water into the fields on the south side of the mesa. What seems to be the plan of the watershed system is pecked into a very large boulder that was once part of the mesa cliffs. The involvement of the entire Tsankawi human community in the creation and maintenance of this water system was intense and is still visible today.

The Acoma people with whom we talked mentioned the community focus on guarding, cleaning, and maintaining the cisterns on top of the Acoma mesa in years past. Imagine the intensity of community intent and practice through song and dance for clouds and rain, which was and is the only source of drinking water in that very arid place. A woman from Taos Pueblo mentioned that the whole irrigation system for the community was set in place centuries ago. And, of course, the location of present-day Taos Pueblo straddling the stream, which flows from the mountains, brings the energy of water into the daily and ceremonial life of the human community. During the winter deer dance, both children and adults are dunked into the icy stream to cleanse body and soul. It is a way of remembering "the source of life." And the source of both water and life goes back to Blue Lake in the mountain above the pueblo. Young and old still walk or ride horses for four days of celebration and connection with their sacred place of origin. The reclaiming of their water place, their watershed, happened in the 1970s and is still considered one of the great legal triumphs of Indian country.

Since all waterways are connected to lakes, hills, and mountains, "it is essential that we spent a lot of time up in the foothills, all around the communities," noted the Tesuque Pueblo man. Visiting these places on an almost daily basis made the entire region intimate. The rocks, animals, and plants were all familiar. How the water flowed from mountains and into valleys was intensely known. Where the fish found harbor and which plants grew alongside the waterways was information needed for survival.

This knowledge was as sacred as the places on the land. Places, water, clouds, plants, and animals—including people—had inherent sacredness in that world. People became capable of healing the land and the water, as both land and water had powers to heal people. In this thinking, respect of every other became an essential element of living. And that respect was shown through prayer, song, and dance. The ceremonial life was important to express acknowledgment of energies that flowed through the land, the sky, and the waters. Prayers and songs are talking with those energies. Dances and shrine visits are an act of becoming one with those energies.

"We live here, we honor water every day. It's in our prayers, it's in our songs. It's in our symbols. All these symbols [used by Pueblo artists] that they have, it's just copying the landscape. The lightning bolt, the rain, the rainbow represent water. Even our hair. When we have ceremonies, women leave their hair long and straight because that represents water. We braid our hair [because] when you pour a pitcher of water into a cup, the water braids itself. It's all connected to water," says the middle-aged man from Tesuque Pueblo.

Life, then, was about continuing to remember the sacredness of the world that we live in, acknowledging our oneness with water, land, and sky, and accepting the sacredness that permeates the wholeness of our place and lives.

Changes

These thoughts and desires encoded in the songs and prayers of the Pueblo world were diminished by contact with the Western European culture. The watersheds surrounding the Pueblo villages were not acknowledged as sustainable systems of living but rather as religious and political domains with artificial boundaries described by those who held human power. Lines were drawn on the ground that had nothing to do with how the wind blew, how the waters flowed, or where the plants grew. Human desires and needs began to separate from and to supersede the native flow of any place. A sense of oneness with Earth, sky, rain, wind, and water lessened with the idea that everything was for the redemption of the individual human soul, human use, or human economic profit.

The religious effort, or missionary colonization, of the Pueblo people by Europeans began during the early years of the seventeenth century, principally along the Río Grande between Taos and Senecu; eastward to the Galisteo basin, the Pecos valley, and the Salinas plains; and westward as far as Zuni and the Hopi towns of Arizona. To the Spanish invaders of the seventeenth century, the economic value of this area was negligible. Its minerals were difficult to exploit, agriculture was limited, and commerce consisted in barter. The only justification for its occupation was as a missionary area. Incredible church and mission structures enclosing volumes much larger than necessary for immediate needs were built. These structures were imposed on the Pueblo village forms for the purpose of becoming the focal centers of those communities. These structures dominated the landscape with the purpose of turning the minds of the Pueblo people toward an abstract heaven away from this earthly place whose character was dictated by sun, wind, and water patterns. Conversion was slow, but there certainly were inroads into the traditional Pueblo way of thinking.

As the political and military branch of the Spanish colonization strengthened, lines in the ground were drawn from the front door of those mission structures defining boundaries of ownership. The Pueblo people were contained within alien boundaries. No longer could communities of Pueblo people move through the valleys and over mesas as they had always done, imitating the water, the clouds, and the wind.

But it was the idea of human ownership of pieces of the Earth that was such a huge, disruptive concept. How could any person own any part of the Earth, the mother, who gave birth to the people? How could any person own part of a flowing entity such as a stream or a river? These continue to be earth-shattering concepts.

Converting people to an alien religious and political system, which defined unfamiliar geopolitical boundaries, was part of the controlling mind-set of the Europeans, who did not talk with the clouds or acknowledge living systems such as watersheds. Intense control of water systems became increasingly important as technologies improved with each succeeding wave of Europeans. Dry farming, which used the water coming from skies and which was the most commonly practiced watering system by the Pueblo people, was too primitive an approach for the Europeans—both Spanish and Anglo. The acequia system of the Spanish colonial period consisted of small waterways that moved water for miles through ground and wooden ditches. They were similar to Pueblo irrigation systems but with horses and metal tools. The acequias became much more extensive than what the Pueblos developed. With the Anglos and greater technology, large cemented irrigation canals and megadams became the standard. Natural watersheds were ignored as more control over the flow of water was practiced.

Large dams were built on rivers such as the Chama and Río Grande. An ex-governor from Cochiti Pueblo bemoans the Cochiti Dam on the Río Grande. "We have the Cochiti Dam right in the middle of our reservation. And it's a good-sized dam. It's the eleventh-largest earth-filled dam in the world. The dam was concluded around 1970. Within a matter of months we noticed that the dam was starting to leak, to seep. In the hillsides you could see the dampness of the land. And so it happened that we just couldn't farm anymore because of the seepage. South of the dam was prime farming land. Families for years and years had fields that were thriving and produced livelihood for the people. But when that [seepage] occurred nothing could be farmed anymore. Water, after a while, was sitting a foot deep, and you couldn't go in with a horse and plow or a tractor. And so farming came to a halt."

He continued, "So that happened and, of course, many tribes faulted Cochiti for it. Actually Cochiti never did anything that wasn't supposed to be done, but we were strapped with the fact that we created this situation, this monster of a dam. But the other thing that happened was that they had to close the river down [to try to repair the seepage]. That was never to have been done either because the water in our area, our culture, is the lifeline to the Pueblo way of life. And to cut off the very flow of water that ties us to our [life] journey is likened to putting a tourniquet on yourself.

"But there's more to the story. One of the things that we learned after the dam was completed was that a sacred site had been severely violated, damaged, right at the site of the outlet. There's a mountain of rock there. It's a good-sized rock that was a sacred site, not just to Cochitis but to Zunis and Santo Domingoans and to many other Pueblos. The rock was shaved in half, cut in half, and it was used to anchor the outlet, the gushing waters from the outlet. The Cochiti tribe never was informed about it. So some very god-awful stuff occurred when the dam was being built. And some people say in Cochiti that it was because of the sacrilegious acts that occurred—the shaving of that monument there, the sacred mound, that as well as the cutting off the water—that we were paying the price of what was happening with the seepage of the dam."

He continued, "The other thing that came with the dam, besides the damage of the fields and the loss of a whole generation of farmers and never being able to get it right again, is the tourism. We all thought that the dam was going to be flood control, but we also learned that it was a recreation site, a recreation dam. There was the tourism that came, but most of them were carrying a six-pack or two of beer and other harder stuff. What it has caused the Pueblo is just a tremendous amount of headaches. A host of problems came with the dam, including the little town of Cochiti Lake."

Downstream, people in San Felipe were feeling the effects of the dam as well. A middle-aged couple from San Felipe remembered their childhood years in that pueblo when they used to travel five miles to their fields. "We could travel along the river in our little wagon. We would see sand turtles, gigantic sand turtles, sunning themselves on the sand dunes. We used to go fishing to get sardine-type fish. We'd go to the shallow waters and we'd make our own nets. We'd get those little fish. I don't think we have those kind of fish anymore. Because of the [Cochiti] dam we don't have the actual flow anymore. It's all being controlled. I'm sure the river snake kind of got mad at that. The water right now just flows straight through. It doesn't have those curves, nice curves, that it used to have with little islands in between. It doesn't take its natural route anymore. In a lot of places they reinforce the sides so it doesn't erode out to the side. But who's to stop the water? Let it go where it goes. It looks unreal, unnatural, to have that steel sticking out on the side of the riverbed. I don't agree with what they are doing."

A seven-term ex-governor of Laguna Pueblo told a similar story about uranium mining on that reservation. The Jackpile Mine in Paguate on the Laguna Reservation was in operation from 1952 to 1982 and hired some eight hundred people. "Unfortunately, with the coming of the mine, people decided to give up farming and instead go to work at the mine. We turned from an agrarian society to one that is dependent upon a wage economy. The transition began to take place in the early 1930s, and by the mid-1960s there was very limited farming. As the mine developed in size and began to move closer and closer to the village, some of our natural springs began to dry up. A number of those springs are really shrines and sacred sites. It's only been since the closure of the mine and the reclamation that has taken place that at least one or two of those springs have begun to show some signs of life again."

Farming in the larger Laguna Reservation was impacted by increased general population in the area and the impoundment of the Río San José into reservoirs and man-made lakes. The Laguna man lamented that "there is a lack of sufficient irrigation waters because of upstream users, the impoundment of the water at Bluewater Lake, the city of Grants, and our neighbors upstream, the Acomas, who had a reservoir built in the 1930s."

In the meantime, a group of three people, one woman and two men from Acoma, talked about the Río San José and the issues in their community upstream from Laguna. "The short, probably no more than ten miles long, life's reach of the Río San José emerges from springs on the western edge of the Acoma grant. Originally, at least a third of the flow came from the Zuni Mountains, and another third came from a major spring located southwest of Acoma, the Ojo de Gallo spring. Today we are down to our final third contributor, the series of springs which are within the river bed itself. Upstream, fifteen miles away, are the municipal communities of Bluewater Village, Milan, and Grants. So during the heyday of uranium mining and activity, huge amounts of people populated those communities. The wastewater of those communities was dumped in the streambed and eventually reached those springs."

One of these three Acoma people commented that they are almost at crisis level. "We don't have baseline flows in the river today. They are so low. And because of that, some of the constituents in the river, like total dissolved solids and sulfates, have formed a crust. They're more pronounced because we don't have the dilution factor. We're at the lowest levels ever in the history of Acoma for the Río San José flows. It's amazing that our little stretch of the river at Acoma is still alive today. It's dead upstream, there's no flow. And there's really no perennial flow downstream. Laguna is even worse off than we are because they're downstream. It's just alive at Acoma, barely alive, barely alive.

"There was a water quality lawsuit by the Pueblo, which was settled out of court. Grants was forced to go to a zero-discharge situation where no more wastewater could be dumped in the Río San José. But there were already decades of pollution in the river. That destroyed the stream. Our people felt that it ruined their agriculture because if you irrigate with that water, you'll see crustation of the soil and the vegetables are stunted. Some people think that so much sludge was deposited into the riverbed and that it's never been cleaned up. Now it's being threatened by new uranium mining, dewatering on a very large scale, a massive scale, and our river can't survive it a second time."

The river that they "could actually swim in, fish from, and just walk along the banks" as children is now out of bounds. Already fifty years ago, parents were "telling us children, 'Don't go swimming in the river.' Many of the children would go over there and come back all white."

These people from Acoma also talked about the connection between groundwater and surface water. "The groundwater is something we can't see because it's not visible on the surface except through spring flows. Groundwater levels in this region below Mount Taylor have gone down over the last fifty years. We have to be very careful with our groundwater pumping that we use to supply our community water system today."

Groundwater in the form of spring flow, along with rainwater, was commonly used in all the Pueblo communities for drinking. Through the years, springs that were used for centuries dried because of regional overuse of water or because they were rerouted. According to an elderly man from Santa Clara Pueblo, "The spring which supplied the Pueblo's drinking water was half a mile away up the canyon. It was first piped right before World

War II and put in a storage tank above our area here. And that storage tank served as pressure to feed the homes. And they piped it through the pueblo into three windmills, two here in the pueblo and then one behind the day school. Now, in the wintertime, when everything would freeze over, the windmills would cease to operate; everyone would go to the creek to get their drinking water. In them days the creek water was clean. Anyway, it was a BIA [Bureau of Indian Affairs] project. It was sort of like a pilot program because they were trying to get the Pueblos into bettering themselves as far as drinking water."

Although surface water from the stream that comes down the Santa Clara Canyon is used for irrigating the fields, three wells using groundwater have been dug in the last fifteen years for drinking and household use. The elderly man said that his community "was fortunate in having a mountain stream and the Río Grande River. And the waters were so different. The river water was real slow compared to the stream water. It was the best thing. You could be irrigating, and even where the water was traveling, if you'd get up close and stepped there, your foot would go down into the dirt because the water was soaking so well as it came along. Maintaining that big ditch was another thing. It used to be a community effort. There was always one man in charge, and he would go ahead and with his shovel he would mark off a portion that each individual was supposed to clean up. It was a big job—making that ditch. They dug the ditch and even went deep with shovels to clean out all the dirt that needed to be cleaned out. In today's way of cleaning the ditch, they just sort of sweep the ditch. Then in later years, the BIA came in to line the big ditch with cement."

The woman from San Felipe talked about how men in that village would "clean the ditch for one whole week. And now it takes them one day. One day, because they don't dig in the way the old people used to. They would dig in and get rid of the mud that settled and get back into the sand. Today, there's a lot of grass and junk along the ditch line and a lot of that moss, that green stuff—algae. It's not a very healthy water system to irrigate your fields."

She elaborated on the fact that everyone in the region does not "realize the development around our communities, like Algodones, all the new houses coming up. They don't just come up. They have to survive, and what do they need, the main thing? Water, water. And, Placitas up on the hill, they pump a lot of water out of our water basin because we have that big old tub of water underneath our valley. Our water level is going down because of all the development. Look at Río Rancho up in the hills there. So many houses coming up. And, our own pueblo, we use a lot of water, because every household has bathrooms now. Even washing machines and dishwashers. A long time ago, we didn't have plumbing; we would have a big bucket right in the house because we'd go get water from the river. Women would go down and do their laundry. They would take their tubs and fill it up, and men would build a fire on the side and boil water right along the river. Of course we kids would be swimming or fishing. Everything was right there—food, water, laundromat.

"We used to drink that water from the river. As far as drinking water now, we really don't drink our water even from our faucet. We buy bottled water."

A woman from San Juan Pueblo said that her father would tell them not to "be throwing stuff in the water because that's where we drink water. It is sacred. Now it's just water. They throw everything in the water and you can't drink it anymore."

The pollution of drinking water is a common issue in the pueblos. A San Ildefonso interviewee talked about "springs that were just popping out from different locations. As our population grows here in this area, water is not coming to the surface as much. Lower Los Alamos Canyon used to have an annual flow. When Los Alamos started sucking up more water, that dried down. Any water that is there in the canyons is polluted with the stuff they put in."

This ex-governor of San Ildefonso said, "We've come such a long ways that we're complacent. How people see water in the ceremonies that we need to continue or that were handed down by our ancestors are going on the wayside, and very few people are practicing or getting involved in the ceremonies to re-create or to bring them back to life. We don't even go out to the four main mountain peaks asking them to continue to provide water. I think we've gotten to the point where we've become so complacent that we go to the sink and turn it on rather than going up to the mountains, the peaks, where the streams are and bringing some of that water home to mix with our waters so that they combine. A lot of people don't pray to water before they get into bathe or to shower or to even drink. Nobody goes to the river anymore except for after the corn dances. Now they're telling us here at San Ildefonso that the fish in the water between the Totavi Bridge and Black Mesa are bad, and we shouldn't eat the fish. I've known since the early years that the fish in the Río Grande are highly contaminated because of Los Alamos, and they're not telling us what the contaminants are. Just think how bad they are in Cochiti Lake. Sometime in the mid-80s, the *Houston Chronicle* wrote about the babies that were born with just brain stems, and they were blaming the middle Río Grande. We were in a pretty bad drought, and Texas was demanding release of waters from not only Abiqui and Cochiti but other dams on the Río Grande. We knew that Cochiti was highly contaminated by chemicals and radiation from Los Alamos. I think that created the birth defects that these children had. How come it's only happened on the stretch of the Río Grande that these people are drinking the water from and watering their plants? Why aren't we doing anything?"

An elder from San Juan Pueblo was also concerned that nobody is really voicing concern for what is happening with farmlands and with our waters. "Nobody is really getting angry about what is happening to our connection to the land. Our young people have no real experience in farming and being in tune with Mother Nature."

The middle-aged man from Tesuque Pueblo believes that our ways are eroding because "we want to not work hard anymore. We like the easy way of life, and working the land is not easy. It's very hard work. And we've gotten away from the belief in the sacredness of water. Water is now looked upon as a commodity sold to the highest bidder. Water now flows uphill towards money. We dance for rain, we sing for rain, our prayers are answered so it rains all day, all night. But we've built very efficient means to get rid of the water, i.e., by lining the drainages, the arroyos, lining the irrigation systems. The ground is no longer used as a sponge. So what we've done is we get into our trucks, put it in four wheel, or nowadays we have those all-terrain vehicles. They drive anywhere and everywhere, so they're destroying the land. We're destructors because every time it does rain, the sediment just flows downhill, and there's now a huge arroyo."

The concern about humans being destructive creatures is not just with Pueblo people. Terry Tempest Williams in a recent interview on PBS asked, "Where is our outrage?" She was talking about the 2010 Gulf oil spill as a regional catastrophe that spread into the coastal areas, into the oceans, and into the rest of the world. She mentioned that the only interest by the news world is assurance that the leak has been plugged, but there is an almost complete ignoring of the destruction of those waters and of the consequences for all life forms. She described with great sadness seeing a group of dolphins watching the oil being burned off the surface of the Gulf waters.

Back in Santa Clara, an older man described, with sadness but without anger, that "the Río Grande used to be pure and clean. We would harvest a tremendous bounty off the river. All kinds of fish—from eels to catfish to carp. One day the U.S. Corps of Engineers came, and they straightened the river out and the river was destroyed. They destroyed a clean and pure system of water clear from its headwaters down through the village. So the animals felt it. All the animals. The numerous kinds of wildlife that lived there that supplied us with food. The redwing blackbird was important to us. But when they straightened the river out, we no longer had that bird. Small things like bamboo that we could use for arrow shafts, we don't have anymore. Nor do we have the eel or the weasel that turns white in the winter."

A woman from Tesuque Pueblo despairingly stated that "if we destroy the land, we're going to destroy ourselves. It's like we're going towards a cliff, and we're all going to end up falling off the cliff."

Hope and Continuity

This woman who predicted doom for our world if we continue "being greedy and consuming left and right and taking beyond our needs" also maintained that "we have to be hopeful." She, like all the other Pueblo people we talked with, maintained that the traditional ways have power, power to heal and to re-create. The meaningfulness of the old ways has not changed, she noted, but we need to remember and practice them—and teach the young people their significance. And there certainly are admirable efforts in the Pueblo communities to give continuity to the coming generations.

A woman from Taos Pueblo described a heritage class that she has taught for five years to young people in that community. "When we first introduced the garden to the kids, they had no concept because every contemporary child turns on the faucet, and there's the water. They don't know what it was like to be a fully operating agrarian society as our grandparents had. The kids didn't know where water came from or that there is a ditch, that there is the whole irrigation system set in place centuries ago. We would talk to them about all the elements that affect our world, both contemporary and traditional, and the natural forces that are in effect. And they saw it too because in our garden when we couldn't get water, they saw that there was trouble with ditches. Our irrigation water comes about a mile and a half off the river. We didn't just teach our kids, oh, here's the water. Go irrigate. They come with us, and we clean our ditches, and we talk about how the water flows and how our pueblo has a direct source in the mountains but that it also feeds all the little veins and arteries through this whole county, and we have to learn to feed off of that.

"We've taken the kids on a river walk and pulled junk out of the river. The first time they came, they turned over rocks and identified all the little bugs that were on the rocks, and we showed them that if you don't have these bugs in your river, that means that something is definitely wrong with the water. We took water samples back to the office, and the kids got to look at them under the microscope. They were able to see the organisms that were visible to your eyes, and they were also able to look at water that contained life.

"Then, we went to the beautiful beaver dam upriver. We talked about what the beaver does. Even though he's taking some parts of the land away, he made us a big old Olympic swimming pool. That beaver family has provided a setting that the kids can learn from and enjoy. If there are natural diversions like that, the water is reserved. It'll eventually go down, but I think that's important for those beavers to have those dams through the water system to take care of fish and provide an environment for us."

The environment, or the community of rivers, rocks, beavers, bugs, and children, was expressed as the most important thing about Pueblo life by this Taos person. "We all have to remember we are here because we are to take care of such places. To bring the rain, to sing the songs. We're not just existing. We realize that from the beginning of time, that's where the ceremonies came from because our ancestors really lived in this place. And, without that sense of community, without that respect for the land, the water and the sky, we couldn't realize the paradise that we live in. We're in paradise. Our grandpa used to tell us that everything we need, everything we need to live here, is here. He also said that we need to get our children into their place in this community to preserve what our ancestors preserved for us—life without all this technology, without all this bureaucracy that prevents things from happening, so that we can flow like the river—clean, fresh, and beautiful."

To flow pristine like a virgin river is a wonderful desire. This desire is a dream that in our world seems distant because we know that our rivers, our watersheds, have been dramatically manipulated, controlled, and polluted. The dominant mind-set of our greater society mostly views our natural environment as another object to be used for recreation, political gain, and monetary profit by governments, corporations, and some individuals. The age-old desire of Western man to be above Nature is certainly being put into practice in our technological world.

This, however, makes it more imperative than ever that we respectfully act as one with our community, our watershed, to help bring some balance into our increasingly fragile and stressed natural world. And these Pueblo voices remind us that we do have the capability as humans to respect and empathize with the waters, clouds, and the Earth to preserve the beauty and integrity of ourselves and our world. Even more, if we do acknowledge that we are water, that we are hydrologic systems ourselves, we have the hope of seeing and honoring our oneness with the larger natural systems within which we live and that sustain us.

Lyle Balenquah, photo by Jack Loeffler

Lyle Balenquah

Lyle Balenquah, Hopi, is a member of the Greasewood clan from the Village of Bacavi (Reed Springs) on Third Mesa. He has earned degrees (BA, 1999; MA, 2002) in cultural anthropology and southwestern archaeology from Northern Arizona University. For over ten years he has worked throughout Arizona, New Mexico, Colorado, and Utah as an archaeologist documenting ancestral Hopi settlements and their lifeways. Currently he works as an independent consultant, but his work experience includes time with the National Park Service, the Hopi Tribe, and the Museum of Northern Arizona. He also works as a part-time river guide on the San Juan River and other rivers in the Southwest, combining his professional knowledge and training with personal insights about his ancestral history to provide a unique form of public education. As a farmer, rancher, hiker, and hunter, he is an active participant in the natural and wild realms, where he strives to better understand himself and his relationship to these environments. As a member of the Lore of the Land, he highlights his own experiences and the experiences of others who live, work, and rejuvenate among the deserts, mountains, and watersheds of the Southwest.

Connected by Earth

Metaphors from Hopi Tutskwa

THIS WRITING IS ABOUT MANY IDEAS: CULTURAL AND BIOLOGICAL DIVERSITY, preservation of traditional knowledge, interactions with our natural lands (*Tutskwa*), and above all, remaining connected to them. Some of these ideas are difficult to describe as they are more or less felt and experienced, or are emotional and spiritual in character. Some are easier to describe in words, others more tangible to the touch, and others more visible to the eye. Yet they are all important to the greater understanding of how we as humans, individually and collectively, interact with and impact our natural environments. By *natural environments* I am referring to the basic idea of landscapes—forests, mountains, deserts, rivers, lakes, etc., as well as the plant and animal life that live within them. I also include the *human* environment as well, as I believe that we are a part of the natural environment, and so how we interact with and impact each other is equally important in this discussion.

The concept of remaining "connected" to our natural environments also has different parts to it. One part is strictly physical—putting some aspect of our bodies in touch with the natural world, breathing it in. Another part is spiritual—having a metaphysical experience or feeling between our natural world that is beyond the physical, but is most likely a result of the physical. And yet one other part can be viewed as a conscious connection—a mental exercise in which without physical and spiritual contact, we remain cognizant of the simple fact that we are indeed a part of a larger natural world.

In my opinion, these different connections go hand in hand, and when you experience one, it ultimately leads to another. Yet getting connected is harder than one might think. In fact, how does one even achieve this connection? It takes a lot of effort, through careful observation, trial and error, and a willingness to better oneself, if even for a moment. Through the various experiences I write about here, I hope to show how I and others strive to achieve this connection. These experiences are singular and collective in Nature.

Given that I am a person of Hopi descent, my thoughts and energies will focus on what I have learned and continue to learn in *trying* to maintain myself as a Hopi person, which is not an easy task. It is only my perspective, and I cannot and do not claim to speak for every Hopi. It simply represents some of the experiences I have been fortunate to have in a constantly expanding understanding of my place within my own natural environments. Our collective human understanding must be like a galaxy spiraling out in the universe—ever expanding with illuminating points of consciousness, separated by parts unknown, that we strive to understand. This writing will be like a river, winding its way through different landscapes and concepts to show the importance of remaining connected to the natural world we inhabit and call home.

I have had many teachers who have helped and continue to help me in this understanding. They need not be named, and my journey is not yet over, but for me, this writing represents one way of showing and giving my gratefulness to them. I think they would see it fit to be this way, as this knowledge is not solely theirs or mine but represents the collective history, culture, and spirituality of many who have come before us. So let us begin.

Ang'Kuktota: Placing Our Footprints

In the oral histories of the Hopi, there is a centuries-old story that originates from the Snake Clan about the first "river runner" in the Southwest. This story depicts the adventures of Tiyo ("boy"), who goes to pray at the river every morning and while doing so wonders, "Where does all the water go?" Determined to answer that question, Tiyo sets out with the prayers of his family in a boat carved from a cottonwood tree, encountering new adventures and people along his river journey. Eventually he discovers that the river empties into the Pacific Ocean (*Paatuwa'qatsi*, "Water World") far from his homeland. Much more happens after Tiyo leaves his boat behind, walking along the beach of the ocean, but for the purposes of this chapter, this will suffice to set the tone. No one is really sure where Tiyo first started from, but many believe that he passed through and over the country found along what are now known as the San Juan and Colorado rivers in Utah and Arizona.

I first heard the extended version of this story along with my siblings from one of my uncles from the Snake Clan, who had learned it from his own elders. I was still very young when I heard this story, but it would not be the last about my Hopi ancestors. Many times in these stories there was some lesson to be learned; often that lesson was simply to *remember*. To remember to remain connected to the landscape and the history contained within. These lessons and remembrances were part of a larger set of traditional Hopi knowledge that spanned thousands of years and many more generations of Hopi ancestors. As each generation learned, many times through trial and error, that knowledge passed to the next generation, and it belongs not to one person, but to the collective whole.

During my youth, my family and I often spent part of our summer vacations visiting the ancient homes of our ancestors throughout the Four Corners of the American Southwest. In these travels we visited numerous archaeological sites such as Mesa Verde, Chaco Canyon, Wupatki, Homol'ovi, and many more. At these places I walked among the preserved remnants of my ancestors' homes and learned from my family members that I was a descendant of those who built these monumental structures. I heard stories about the Hisat'sinom (the "Ancient People") who possessed remarkable skill, ingenuity, and determination.

These stories told of people who could grow crops in the driest of climates, who could communicate with supernatural forces, and who could make it rain and snow with the power of their prayers. I also heard stories of great calamities and misfortune that befell my ancestors due to their own greed, corruption of power, and forgetting their spiritual and earthly connection to their natural world. Yet at that early age, I never fully understood what was being said to me. Only in later years would I come to appreciate the depth and complexity of Hopi ancestry.

Fast-forwarding a dozen years, I once again found myself among the places of my ancestors. As a would-be student of anthropology at Northern Arizona University, I worked as an archaeological technician with the National Park Service (NPS) in Flagstaff, Arizona. For the next eight years I studied the science of archaeology and did my best to *be* an archaeologist. A lot of our work included tedious recording and mapping of the architectural spaces, features, and materials of ancient structures found within various national parks and monuments. As important as this work was to the "mission" of the NPS, it was not spiritually fulfilling, yet I continued to remember the stories told to me by my family and did my best to also *be* a Hopi.

Through it all, I wondered who these people were that built these places of mud and stone. During certain tasks, such as documenting original construction mortars found in the pueblo walls, I would notice the preserved fingerprints of a Hisat'sinom builder in the mortar. In these moments, I recalled the stories I was told as a child, and in my mind's eye I could *see* these people as they once were. I would often discuss my job with my family at Hopi; what we did, what we studied. They would listen intently to my descriptions of the features and artifacts we recorded. Again, I would hear stories about our ancient ancestors.

Only this time the details were more profound and clear to me. Being older, I was beginning to understand how we as Hopi people had come to be who we are. How multitudes of ancestral Hopi clans had traveled far and long across the landscape to the place we now called home. I understood that in these travels we learned how to be Hopi. That being Hopi was not a right but a privilege that was hard earned and came at great cost of effort.

I was also now of age where I was an active participant in some of the Hopi ceremonies where I learned that what we did in those rites was reenact our connection to landscapes and places. I came to understand that our ceremonies and rituals had their origins deep in the ancient past of our ancestors when we learned the values that serve as the basis of modern Hopi culture: cooperation, humility, hard work, and stewardship of the lands we live in and rely upon.

I learned from my family members that the artifacts and "ruins" I studied in my archaeology career had deeper meanings beyond what my scientific data sheets could relate. I was told that these objects and places were the "footprints" of our ancestors, and their presence at archaeological sites continues to provide physical and spiritual connections to our past. Leigh Kuwanwisiwma, director of the Hopi Cultural Preservation Office, offers this perspective:

> These "footprints" that are the hallmark of Hopi stewardship over these lands are described as ruins, burials, artifacts, shrines, springs, trails, rock writings [petroglyphs, pictographs, or other forms], and other physical evidence of occupation and use. Thus, archaeological sites [which can be seen as silent reminders of the past] are not mere vestiges; Hopi rites and liturgies recognize them as living entities.[1]

This physical act of leaving our footprints has come to be known in Hopi as *Ang'kuktota* ("along there, leave footprints"). Yet another conceptual way to think of this term is as "ancestral migration paths." This term and the philosophical concepts behind it provide a deeply spiritual perspective for understanding our ancestral past. It also provides a method to explain the Hopi perspective of the archaeological record to the larger non-Hopi audience, most of whom are educated and trained in Western-based scientific paradigms. In a sense, the Hopi way of conceptualizing the ancient past provides a new lens to analyze the work that archaeologists conduct.

One aspect that always struck me about my ancestors was their inherent closeness with their natural environment. Of course they literally lived "out in the wild," but to them, it may not have seemed that far away from anything as they were always right in the middle of it; "it" being the greatest cathedral of them all, the natural world where they interacted with their natural environment on a daily basis. The intimate settings of their homes within their surroundings, high up in the alcoves, deep in canyons and along the tops of sweeping mesas and buttes, created the ultimate awareness of just exactly where they were, physically and spiritually, within the larger world.

In many cases, they lived close to their resource bases of rivers, springs, forests, and hunting and gathering areas. In essence, they lived literally among the resources that sustained them from day to day. This is of course a far cry from today's modern existence, where most people never see their food before it is served to them or arrives at the grocery store. In that sense, many of our humankind have definitely lost some part of the connection to our ancestral lifeways. How many of us continue to view our natural world with eyes that not only see the landscapes, but sense the inherent energies that still exist there? We seem cut off from the greater forces that have shaped our lives for thousands of years.

Our modern-day dwellings are also symbolic of our growing disconnect from our natural surroundings. The sole intent seems to be to keep ourselves sheltered and hidden away from each other and the landscapes just outside our doorsteps. In contrast, the homes of Hopi ancestors represented the closeness with their natural surroundings. We see earthy tones derived from native materials such as quarried rock and flagstone, soil mortar, clay plaster, wooden beams, and vegetation; all of which are products of the Earth, sun, and water. Even the world-renowned architect Frank Lloyd Wright understood this philosophy centuries later, stating, "Buildings, too, are children of earth and sun." The settlements of Hopi ancestors were literally "born" out of the Earth, and this symbolic birth reflected their connection to the environment they lived in and relied upon.

Today when a Hopi person visits such ancient places, we don't simply see the remnants of a bygone era; we see reflections of who we once were and what we have now become. We witness the artistic and technical accomplishments of Hopi ancestors, but we recall the spiritual accomplishments of our ancestors as well. We are reminded that in order for the present generations of Hopi to flourish and prosper, we are dependent upon the gifts of our departed ancestors. T.J. Ferguson and Kuwanwisiwma write the following:

> Ancestral villages that have fallen into ruin are not dead places whose only meaning comes from scientific values. The Hopi ancestors who lived in these villages still spiritually occupy these places, and these ancestors play an integral role in the contemporary Hopi ceremonies that bring rain, fertility, and other blessings for the Hopi people and their neighbors throughout the world. *Itaakuku*—footprints—are thus a part of the living legacy of the ancestors, and they play a vital role in the religious activities essential to the perpetuation of Hopi society.[2]

The idea that the ancestors remain among us in the present is directly based on the Hopi concept that the meaning of the past is what it contributes to life in the present.[3] In essence, by acknowledging our ancestors' existence, they acknowledge ours through the answering of our prayers. This understanding provides a continual connection between modern Hopi people and their ancestors. This connection is contained within the landscapes, wherein Hopi ancestors interacted with their natural environments, leaving a legacy behind that their descendants must now strive to continue.

It is this connection, between ourselves and the world we inhabit, that continually needs to be rebuilt, maintained, and strengthened. This can only be achieved by actually getting "out there" among the wild places so we do not forget how our ancestors remained connected. If we agree that from a Hopi perspective, "history is in the land," then much can be gained by following the footprints of our ancestors out onto the landscapes, where the spirits of their successes and failures still remain for us to learn from. Let us move on into another connective landscape.

Beyond the Hunt: Pursuing Life

I am a hunter. Or to paraphrase sixteenth-century French philosopher, Rene Descartes, "I *hunt* [think], therefore I am." Although Descartes was focused on the philosophical question of how one determines if he or she indeed exists or not, his statement provides the answer. For if one is thinking, then that proves there exists an *I* to do the thinking. The basic premise is that in the act of thinking, I am alive. Therefore, in the act of hunting, I too find that I am alive. But can the act of taking another animal's life really be that complex and "deep"? Well, yes and no.

I have been a hunter for most of my life, as have a large portion of Hopi males. It is something that begins early and, for some, becomes a life-long learning experience full of discoveries about oneself. At the age of five I was first taught to shoot with a homemade slingshot, carved from a simple branch from a cedar tree, slung with recycled bike tubing and an old piece of leather. I learned to find just the right stone to use, how to aim between the *Y* of the sight, and then how to let it fly. Many small birds met their demise with this simple device. I was also learning to shoot with a small bow and arrow, given to me as a gift from the Hopi spiritual deities known as *Katsina* (singular), which visit the Hopi villages starting every spring through the summer. I will not delve into the *Katsinam* (plural) too much as that involves a highly spiritual aspect of Hopi culture that is not necessarily appropriate for this discussion. There are many books and other literature available for those who wish to pursue that subject further.

Basically, the Katsinam visit the Hopi villages, performing dances of prayer and good will. They also bring various gifts for all members of the community, their friends and benefactors. For the young girls and boys they bring elaborately and beautifully decorated gifts. The girls receive dolls, plaques, and other items. The boys usually receive a handmade bow complete with a set of three or four arrows. These are prized possessions, and when we boys received them, we would waste no time in putting them to use—shooting boxes, cans, tires, and maybe even nipping the stray cat that crossed our paths.

However, most of the bows were far too weak to actually kill an animal, and they usually broke in half after some use. But with them, we learned some of the basic skills of hunting: observing "game," stalking, aiming, and basic motor skills. These gifts were also a reflection of the recognized gender roles within Hopi culture. Males are the hunters, with a responsibility to provide for their families with the wild game they bring home. There were also important cultural aspects to the act of hunting, which I will get to later.

After a time, some of us graduated to "heavier" armament like BB guns and small-caliber .22 rifles. That's when "hunting" really became hunting and also began to have more purpose to it. It was at this time that my cousins and I would embark on small hunting expeditions to various areas around our homes and hunt jackrabbits and cottontails. We would bring them home to our grandmother who would ceremonially receive them, a ritual involving the placement of cornmeal upon the rabbit's body and saying a small prayer of thanks for the nourishment it would provide. Afterward she would prepare a small meal where we all ate and told of our adventures from out and about.

This latter aspect, the getting out and about, was just as important as the actual act of hunting. In our "expeditions" we often visited certain places time and time again, learning the intricacies of each place. In doing so, we became familiar with our surrounding landscapes and learned the skills of being self-sufficient, drinking directly from springs and eating some of the game we killed along with wild plants that were known to us. Often we would come across old ruins, trails, or petroglyphs, and in our later discussions with our family members, they would tell us the history of the places we had encountered. Slowly over time, we developed a mental picture of the landscape. It became a part of our larger world and, as a result, maintained an importance in our lives. The concept of remaining connected to our natural environments, physically, spiritually, and consciously, started to take seed in our minds.

Somewhere along the way, we were deemed "ready" by our fathers and uncles, and it became time to really become serious about hunting and all that it entailed. Thus we began pursuing larger game such as deer, antelope, and elk. This was a much more involved process and required a great deal of physical and spiritual preparation for the coming hunt. Sometimes we fasted before our hunt to cleanse ourselves physically and mentally. The reasoning stems from the belief that humans are the "dirty" ones entering a pristine spiritual wilderness. We may bring negative energies with us that affect our hunt and result in disaster. Many times, prayer sticks and offerings were prepared for the game we pursued; these were taken to the hunting areas and deposited along with prayers of success and safety. Hunting, as many know, can be a very dangerous endeavor.

Thus the hunt would begin. However, it wasn't so much about the "kill," although the wild meat a kill provided was definitely appreciated. It is the *process* of hunting that I believe has more importance. It is in this process that one again endeavors to maintain the connection to the natural, wild world. Oftentimes during hunts, I take more pleasure in just being there, seeing the early sun rise, observing animals in their natural habitats, and watching stars revolve around in the clear, cold night sky. I often hunt the same areas year after year, and just like the lands I hunted in my childhood, I have come to "know" the lands I now hunt as an adult. They are a part of me, all of me—mind, heart, and soul.

There is another reason why I hunt. Being out in the wilderness I actually feel more alive, more of a participant in my natural world and the natural cycles that occur there. I once asked my dad, "Why do we hunt?" Sitting on the edge of a mesa, staring out over a landscape dotted with grassy meadows, surrounded by stands of aspen and fir, with cumulus clouds building on the distant horizon, he remained silent while pondering this question from his son. After a while he sighed heavily and spoke: "I guess I hunt because out there is Nature, things living, things dying. I want to be a part of Nature, just as our ancestors were."

Hunting is definitely a large part of our Hopi culture and history. Knowing that my ancestors hunted in some of the same areas I now hunt, I do feel closer to them. I sometimes wonder how much they would recognize if they were standing next to me looking out over the landscape.

In some instances, they are closer than one would think. Once during an elk hunt in northern Arizona, I came across a projectile point (an arrowhead) lying on the ground beneath my feet. Black, shiny obsidian reflected sunlight against the backdrop of white limestone that surrounded it. I had seen numerous and varied types of these points found on archaeological surveys and other journeys across the terrain.

Many of these points had specific uses, the size and shape made for specific game. This type was used to tip the dangerous end of an arrow, and when it was first made, it was sharper than a modern surgeon's steel scalpel. Its existence showed that my ancestors once roamed this same area, pursuing the ancestors of the elk I now chased. Holding the point in my palm, I wondered who the ancient hunter was who dropped it. I could imagine him, clad in buckskin that he had tanned himself, carrying a bow made of oak, strung with twisted sinew that launched sumac arrow shafts tipped by black obsidian. In my mind's eye, I could see this hunter, moving stealthily along in moccasins, peering around tree trunks, silently praying for a successful hunt. In contrast, here I stood. A modern "savage," clothed in synthetic camouflage and armed with the latest technology that was a reflection of the modern world I lived in. Who had it better, I wondered.

Yet another time I stood before a petroglyph panel depicting a classic hunting scene: a hunter with bow drawn and a small line indicating the flight of an arrow toward an antelope. What struck me wasn't so much the scene itself, but the manner in which the figures were represented. The hunter was puny looking, his bow fragile and his arrow struggling to maintain flight. The antelope in contrast was huge, a sure giant of the animal kingdom, towering over the quivering hunter. I thought to myself that this scene depicted only one thing—the truth. Hunting is not easy. It requires great skill, strength, and a whole lot of luck and prayer. The hunter who made this scene knew the score, and he knew it all too well. Yet, like modern-day Hopis fasting and praying for a successful hunt, this hunter, too, left his prayer pecked upon the stone wall. I walked away, wondering if he had killed.

And yes, let's not deviate from the cold, hard truth: hunting is also about killing an animal to eat it (and for some, to use the parts—antlers, bone, hide, sinew, hoof). This is a harsh reality that all hunters must face, and how we choose to face it varies from hunter to hunter. In the case of the hunts my dad and I embark upon, we realize that in the taking of an animal, we are participating in the natural cycle of life and death. We hunt to eat. Some may say that in this day and age, hunting is not necessary. We can simply go to the store and buy all the meat we want and gorge ourselves day and night! Well, where is the experience in that? How does that allow me to connect with the natural, wild world? I say that if I am what I eat, then I would rather eat wild Nature, naturally. In short, filling my belly and the bellies of my family and friends with wild game I bring home from a successful hunt allows us all to remain connected with our Earth.

From a Hopi perspective, bringing back an animal from a successful hunt also involves a whole slew of cultural practices performed on behalf of the animal, both in the field and when it arrives "home." Just as we performed prayers for ourselves and the animals before the hunt, we now do the same after. We recognize the life this animal had and the new life it has now embarked upon. We acknowledge that in order for that animal to exist, it must have someplace wild to roam. In our prayers, we ask that the animals may prosper, that they have a healthy environment in which to live, that in the cycle of life and death, we partake now but will one day give back too.

So, again, is hunting really that complicated or simple? My answer is again, yes and no. Hunting allows me to feed my family in some small way (perhaps the simple answer). Hunting also allows me to remain in contact with the world my ancestors once roamed, and in doing so I maintain my own "wild" connection to the Earth (perhaps the complicated answer). In his book *Heartsblood: Hunting, Spirituality, and Wildness in America*, author David Petersen puts it another way:

> In the end, we find sacredness only where we seek it. And only *if* we seek it. Authentic hunters, nature hunters, spiritual hunters, seek

and find sacredness in aspen grove and piney wood; in mountain meadow and brushy bottom; in cold clear water and stinking wallow; and ultimately—necessarily, naturally—in bloodstained blood.[4]

Riding the Serpent: Following in Tiyo's Footsteps

In Tiyo's time, as is the case today for modern Hopis, rivers, lakes, springs, and other bodies of water were home to a great natural force: the Water Serpent. Indeed the image of the Snake or Feathered Serpent is a prevalent image and concept in many cultures throughout North, Central, and South America (an indication of perhaps common ancestry or relations deep in the past). As children, we were told never to play near a spring around the middle of the day as that was when *Paa'lÖlÖkong*, the Water Serpent, would poke his head out. If you happened to be around, he just might take you with him back into his underwater home. (Fortunately that never happened to anyone I knew, but some swear they have indeed seen him poke his head out, like some kind of Hopi Loch Ness Monster.) Still, it was a powerful method of teaching us kids about the immense physical and spiritual energy that water contains.

So it is with the oral tradition of Tiyo's journey. In that distant age, the people may not have known exactly where the rivers originated from or where they ended up. But for sure, they could sense that these waters were bringing great strength and spirituality with them. As the people prayed to the spirits of the rivers and other bodies of water, those prayers were carried to great forces, hopefully to be answered one day. Even today, Hopi people are taught that when they encounter bodies of water, be it a river, a lake, or a spring, they take some handfuls of water and symbolically "throw" that water back to Hopiland so that the rains and snows will come to the dry landscapes of the Hopi mesas. In essence, water, in all its forms, was a resource to be appreciated, respected, and never abused.

In Tiyo's case, he was at least the first Hopi who braved the power of the river and lived to tell about it, but not without many close calls. In most instances Hopi ancestors were usually found *along* the river, as opposed to *in* the river. Numerous ancestral Hopi villages and settlements are located along the great rivers of the Southwest, and they continue to be honored in story, song, and prayer. Some of the Hopi names include *Pisis'vayu*, an archaic term referring to the Colorado River; *Yotse'vayu*, the Ute River (the San Juan); *Hopaqvayu*, the "River of the Northeast" (the Río Grande); *Hotsikvayu*, the "Winding River" (the Verde River); and *Palavayu*, the "Red

River" (the Little Colorado). As attested to by these names and meanings, these rivers and many others continue to remain a viable part of the Hopi cultural landscape and serve to connect modern Hopi people to regions located far from the current Hopi Reservation.

Yet while these waters remain culturally important to the modern Hopi, historically there was little consideration of this continued importance to the Hopi and other tribes by modern politics and federal guidelines. Many decisions are made by politicians on how rivers in the Southwest are to be managed and used, but most, if not all, of these decisions do not address the interests and needs (let alone the cultural relevance) of rivers to Native tribes, including Hopi. However, there are some renewed attempts by the federal government to include perspectives of Native tribes, particularly the Hopi, in current management strategies of resources in and along southwestern rivers, including one of the greatest rivers in the entire world, the Colorado River, as it runs through the Grand Canyon (yet another important cultural landmark for Hopi).

Throughout the 1990s, the Hopi Tribe was involved in two research and documentation projects concerning the Colorado River and the Grand Canyon. During the initial years of 1991 to 1995, the Hopi Tribe became among the first Native American tribes to request "cooperating agency" status in the development of the Glen Canyon Environmental Studies (GCES), which resulted in a comprehensive overview of Hopi history and culture related to the Grand Canyon.[5] In subsequent years, 1998–1999, the Hopi Tribe was again a "cooperating agency" in the development of the Glen Canyon Dam Environmental Impact Study (GCDEIS), a lengthy documentation and research project undertaken to assess the impact of the operations of Glen Canyon Dam on the natural and cultural resources found along the river corridor. The work the Hopi conducted on GCDEIS built on the previous GCES and resulted in another report specifically documenting Hopi ethnobotany perspectives and information.[6]

Both studies were parts of a larger undertaking entitled the Glen Canyon Dam Adaptive Management Program (GCDAMP), administered by the Grand Canyon Monitoring and Research Center (GCMRC), an entity of the United States Geologic Survey (USGS). Funding for both of the studies originated with the Bureau of Reclamation (BOR), which operates the release of water from Glen Canyon Dam.

The work the Hopi groups conducted during these two projects was successful in showing the vast and complex set of knowledge that Hopi people still retain about a region that is located well outside modern reservation boundaries. But let's be honest and say that political boundaries, such as the reservation, are quite arbitrary and meaningless for most Hopis. Our connections to lands have no boundaries, just as our knowledge about these places traverses boundaries and wipes them off the map. The idea of a mental cultural landscape remains within traditional Hopi knowledge.

During these cultural trips, Hopi "researchers" (i.e., knowledgeable Hopi people representing clans, religious societies, herbalists, artists, and farmers) spent considerable time documenting Hopi perspectives concerning cultural and natural resources found along the inner river corridor. Documentation came from various river trips, five in the first study and two in the second study, which were guided by Anglo river guides and other scientists from various agencies who were familiar with the logistics of getting to and from these sites. That isn't to say that the Hopis didn't have a lot to say about the places. Many of the Hopis who were included in these trips had indeed heard of these places through the oral tradition as passed down from their own elders.

Thus they came with a wealth of cultural knowledge that helped to bring the Hopi presence within the Grand Canyon from the past (a static archaeological perspective) into the modern era where the culture is anything but static. Hopis have always stated that we are a *living* culture. That is, the knowledge about our history isn't relegated to just the past; it lives in the present among the Hopis who retain and continue to use such information in our daily and ceremonial lives. Whereas strict archaeological perspectives portray ancestral Hopi lifeways as relegated to the "prehistoric," Hopis view our lifeways as a continuation over time, constantly evolving with our interactions within our environments.

As a part of the Hopi research, hundreds of ancestral Hopi sites and even hundreds more plants and animals that hold central roles in modern Hopi culture were documented. So it came as no surprise to the Hopi groups that these plants and animals were also found during the archaeological work conducted along the river. It proved that our knowledge of the natural world has traversed time, carrying on from one generation to the next, the concept of the *living* culture of Hopi shining brightly in the archaeologists' excavation pits, yet more importantly, within the minds of the modern Hopi people.

Hopi people always consider our ancestors as among the first to experience the spirit of the Canyon. Not only is the Grand Canyon an origin point for many modern Hopi clans, but in historical times, certain members made pilgrimages down into the depths to collect resources and deposit prayer offerings at selected sites. These pilgrimages and offerings are a symbolic remembrance and continuation of our long and complex history within the Canyon, which has never been lost by the Hopi people. The research conducted by the Hopi Tribe and its members has solidified what we have always known: we *are* the Canyon.

A more recent development in the Hopi presence *on* rivers is the pursuit of river guiding by Hopi tribal members. Through various means and opportunities (too tedious to fully describe here) some of us have begun to learn the skills of what it takes to take a raft down the river. Some of us who were once only passengers on the boat are now behind the "sticks" and finding out firsthand what it takes to be a boatman.

A handful of us are now working as part-time guides on rivers in Utah and Arizona (including the ones that Tiyo first journeyed upon). For me personally, these experiences have given me a new way to remain connected to the physical and spiritual aspects of the natural landscapes. When one is floating upon the water there is no doubt that you are engaging in a spiritual interaction with forces greater than you. Whether it is flowing over massive waves on the Colorado River in the Grand Canyon or drifting lazily along on the current of the San Juan River, there is something very therapeutic about just *being* there. As you float past towering canyon walls, enjoying the concert of early morning birds or watching the full moon emerge over a distant horizon, one cannot (I would hope) not be forever changed by those experiences. It leaves one with a greater appreciation of the diversity found in Nature.

Often, some of the greatest interactions aren't with Nature at all but are with the passengers who accompany us on these trips. For some, including many youth, it is their first real encounter with the wilds of Nature. Many have never slept out under the stars or witnessed animals such as beaver, bighorn sheep, or a pack of coyotes howling at the moon. I often talk with them about the human need to have these experiences. Somewhere deep in our psyche, we still retain that sense of what the wild is, the wild that our ancestors all experienced.

For a handful of days out on the river, we are given the opportunity to rebuild and maintain our connection to the natural world. It is during those days that I often witness the spiritual awakening of people who may have never thought of themselves as being spiritual. Yet, by the third or fourth day of hiking, paddling as a coordinated group through waves, sleeping on the sands of the beach, they suddenly feel as if they had been doing this all of their lives. I believe, again, in the spirit of Edward Abbey and others, it is because we have never fully been able to break free from the adventurous ways of our ancestors. Oh, indeed we have tried, and it gets easier to forget that long history with every new passing technological breakthrough.

But try as they may, most of our passengers give in to the energy and find themselves looking at the world through a new set of lenses. They finally see what they could not see before, and it is that awakening that I wish more of today's society could come to appreciate. Suddenly they are remarking that if given the chance they would not return to the confines of civilization back at home. I can only smile at such statements, knowing that we all have to return to the man-made world, but perhaps because of this experience they will possess more compassionate and wiser attitudes toward the world we are charged with caring for.

Often I receive correspondence from some of our passengers after the trip is over, and they tell me their outlook on life and the world in general has changed. They are aware of the need to preserve and protect our wilderness areas from ever-expanding development. Yet they often ask how one goes about actually doing the "preserving and protecting." A valid question indeed and one that does not have an easy answer.

I suggest that they consider donating funds or volunteering time to a land stewardship group (but to do their research and find one that has similar values to theirs). I tell them to vote for political candidates that support land stewardship issues (again after careful research). But the most important thing I suggest, which is also the simplest to do, is to remain active in the outdoors. Continue to get out and experience it firsthand, as often as possible. For this I believe is crucial in rebuilding, maintaining, and strengthening our collective connections to the natural world. Again, author David Petersen provides us with another perspective: "In wild nature I find spiritual solace and a cathartic reaffirmation of cosmic sanity that I find nowhere in the made world. To the contrary precisely: Nature is the only antidote to civilization."[7] Amen.

Prayers into the Earth: Lessons from the Cornfield

Returning back to Hopiland, I also find that "cathartic reaffirmation" Mr. Petersen speaks of in the roles of just being Hopi. One of our Sisyphean tasks is to try to be farmers in one of the most arid regions in the world, where we receive a meager eight to twelve inches of moisture per year in the form of summer rains and winter snows. Granted, we are pretty good at it and have been successful in developing strains of corn and other crops that are adapted to our dry climate. Yet this process can also build great humility. Still, year after year, we return to our farmlands and plant not only seeds but our prayers into the Earth for a successful harvest. And then the real work begins.

There are many teachings and skills associated with being a traditional dry farmer, and each person will learn from those closest to him or her. My main teachers have been my father and uncles from my paternal side of the family, and it is on their fields (which are the property of the female clan members) that I have received my greatest understandings of what it means to be a Hopi. Starting at an early age (much like hunting), we are aroused from slumber in the predawn hours and loaded into the trucks, headed out to the cornfields to begin our "lessons."

Hopi farming, for the most part, is dry farming. That is, we do not irrigate our fields with constructed canals that bring water from a stream or pond. All the moisture the corn plants will receive comes from that already stored in the soil from the previous winter snow pack or the coming summer monsoons that provide direct watering or produce runoff that runs through and over the fields. The planting process is one of routine that after a time becomes second nature. Traditionally it is also all done by hand with no machinery (although a tractor may be used to clear the fields before planting begins and to routinely "weed" between the rows once the corn becomes established). As with a lot of Hopi practices, it is in the process of actually doing something that one learns the most.

There are many metaphors and lessons contained within the work of being a Hopi farmer. In the act of planting the seed, one learns the importance of perseverance. The process is straightforward: kneel, break through the hard crust into the moist soil beneath with a planting stick (traditionally made of oak or other hardwood but now often a metal pipe flattened at one end), and dig a hole reaching a depth of eight inches or more depending on soil moisture content. Then, take ten to twelve seeds in your palm and deposit them into the hole (the seeds' "home"), take the soil you excavated and carefully rub it between your palms to break up the hard parts, and return the moist soil back into the hole. Finally, cover up the hole with dry soil to ensure that the soil beneath does not dry out. Repeat until finished. Depending on the size of the field, number of fields, and how much help you have, this may take anywhere from a day to a week.

To the outside world (i.e., non-Hopi), this may seem inefficient, but it has a deeper meaning that relates to the concept of remaining connected to our natural landscapes. Though the process is backbreaking and monotonous, it reminds the farmer of one of the Hopi tenets: hard work is necessary to achieve something good for one's family. This leads to another Hopi truth: farming requires one to have a healthy and positive caregiver connection with his family and children, both human and plant forms. When Hopi corn seed is put into the ground and subsequently sprouts, it is a literal representation of having children. Just like human children, your corn children will require and demand constant care and attention from you. Of course, nowadays most Hopi men need to work wage-paying jobs during the week, so this leaves the corn vulnerable during the day. It also used to be that young Hopi boys would be put on guard duty day and night to chase any critters away, but this is also rare now (both of these conditions being a result of the needs required by living in the modern world).

So the onslaught begins. First, the mice will pay a visit and dig up the seeds and eat them. Next, the crows will systematically walk the cornrows eating what the mice didn't eat. If you're lucky, you'll get a few to sprout, but here come the cutworms, who will eat them from the bottom up. Whatever is left will be subjected to hot, dry days with little or no rain. When it does rain, it will surely rain on the other guy's fields just over the hill. Or it will rain so much that a flood will come through and simply wipe out everything. That's humility.

But eventually your perseverance will pay off, and on cool summer evenings as the rains fall soft and steady on your field, you can stand among your children and listen to the winds blow through the leaves. A few months later, you and your extended family can fill metal washtubs and buckets with ears of corn colored blue, red, purple, white, and speckled. In the distant past, this meant that you probably wouldn't starve that year. In today's modern era, it means that you get to do it all again next year. In the meantime, your female family members will expertly clean and store the corn (which becomes their property) and will cook and prepare numerous dishes with it.

One of the biggest contributions of being a Hopi farmer is in maintaining unique corn varieties that cannot be found in any other part of the world. For generations, Hopi corn seed has been a greatly held secret, but over the past century small samples have been collected and studied by Western scientists to decipher its properties. Some Hopi people fear that this will result in a loss of intellectual property that they and their ancestors have developed over thousands of years. They fear that Hopi corn seed will become a "brand name" product, offered up for sale by private corporations who only seek a profit. This is a valid concern in today's world of privatization, monopolization, and exploitation of food crops by megacorporations.

A few nonprofits and other organizations do currently offer small quantities of Hopi corn seed (and other Hopi seed crops) for individual purchase, but their activities do not seem to pose a threat to the Hopi farmers, at least not at the moment. It is said by Hopi farmers that Hopi corn should never be sold for profit, that it was given to and chosen by Hopi ancestors to be the birthright for future generations of Hopi people. So despite intrusions into a traditional way of life by modern economics, and facing the potential loss of our hard-earned intellectual property to outside interests, the largest concentration of Hopi corn seed, and the knowledge needed to properly grow and care for it, still resides with the Hopi farmers and their families.

Cultural and plant diversity is very important to the overall scheme of the natural world. What works in one region may not work in another. I was once part of a group of Hopi students who traveled to Creel, Chihuahua, in northern Mexico in 1999. We traveled there to attend a language conference, but we also learned that the Raramuri (Tarahumara) Natives who lived in the region were suffering an extended drought resulting in a loss of much of their corn crops for that year. When my uncle heard of this, he gave me some large bags of Hopi white corn seed to take with me (Hopi white and blue corn are considered to be the hardiest of the Hopi corns). His intentions were to help out the farmers and to see if the Hopi corn would grow in a different part of the world. We arrived and presented the seeds to some local farmers, who proceeded to eat most of it, but eventually they did attempt to plant it in their fields. The word we received from a local anthropologist (there's always one of those in Native communities, isn't there?) is that the Hopi corn did germinate, and some of it reached its proper height of only three or four feet, much to the amusement of the Tarahumara farmers who were used to corn plants much taller. The short height of Hopi corn is one of its genetic specialties for surviving in the harsh desert. Most of the Hopi corn plant is not above ground but below the surface where a long taproot will extend far into the soil to capture any deeply stored moisture. However, the climate changed and it rained—a lot. Eventually most of the Hopi corn plants literally drowned and rotted away in the much wetter soil. Our experiment to introduce Hopi corn to another part of the world had failed, but it also proved the Hopi point that the corn we grow was destined for us and us alone.

Traditions state that the Hopi way of life would be characterized by three items: corn seed, a planting stick, and a gourd of water. These items would teach and remind us of the important values of Hopi culture: cooperation, humility, hard work, and stewardship. These three items, simple as they may seem, would provide the foundation for maintaining our connection between our ancestral past and the present day we now inhabit. This history, this connection, is ultimately recorded upon the landscapes our ancestors once traversed and continue to care for. Mr. Vernon Masayesva, former chairman of the Hopi Tribe and now director of Black Mesa Trust, states the following:

> Early in life . . . when we are taught to plant, the elders would tell you that if you want to plant a straight row of corn, you have to first pick where you are going to be going, where you wish to end up at. And then you start planting, but every so often you have to look back. Because it is what happened that tells you where you are at, and where you are going. And this is why cultural preservation . . . is very important. You never know where you are going unless you understand where you have been.[8]

Creating a Commons: Finding Our Balance

So just what exactly was all this about? Was it simply a chance for me to relate how lucky I and others have been to participate in these experiences? Was it an opportunity for me to say, "Hey, look how great Hopi culture is!"? Or was there something else behind it all? I leave that up to you to decide and take from it what you will.

It is time to take off the microlens and zoom out to the larger picture. This project sought, in part, to bring attention to the ways in which we, as humankind, interact with our cultural, natural, and spiritual environments. The purpose is to illuminate how our actions and interactions with these environments result in consequences.

My goal was to illustrate how one slice of the global population, in my case Hopi, has attempted to achieve positive interactions with our environment. We too have not always been successful at achieving that goal, as our history attests to the many failures our ancestors made along the way, and we will undoubtedly have more. Learning in Hopi culture is always best achieved by actually getting out there and doing it, whatever *it* may be. Within us all we have the opportunity to do something *positive* for ourselves, our children, our families, and ultimately for the natural world we depend on.

In the not-too-distant past, it seemed that we were all isolated from each other, living our separate lifeways and traditions. The other side of the landscape, the other side of the world for that matter, was a very far-away place that probably was beyond our concerns. What happened over the hills and through the woods was the problem of those who lived there, not ours. However, in today's modern age, that isn't the case anymore. We aren't as isolated as we think. One only need flick on the radio, television, or Internet, and the other side of the world suddenly is right in our own backyard. It's hard to say that we are no longer affected by others' problems.

For hundreds of years, since the first arrival of Europeans in the Western Hemisphere, our "isolation" kept outsiders at bay from the Hopi landscape, our homes. Yet eventually they came and kept coming, so that no longer were Hopis isolated. The changes to Hopi culture wrought from this de-isolation have been both beneficial and detrimental. The spaces between us are growing smaller, shrinking away every day so that we are now forced to interact with one another, for better or worse.

Protecting the spaces, the *natural* spaces, between ourselves now isn't just others' problem; it becomes ours. The boundaries on the map are becoming blurry, forcing us to rethink how we view and define our landscapes. In rethinking these boundaries, we must also rethink how we address the issues and problems facing our shared natural spaces. We must begin to think of these spaces no longer as dividers, keeping us apart, but as common grounds, being impacted by all of our actions and intrusions. It is a hard endeavor but one that must be done in order for us to survive.

Mr. Jack Loeffler states that part of the work of this project was to create an idea of the "commons." Through this project, working with my fellow colleagues and Hopis, I have come to understand that a *commons* is a collective understanding about not just physical lands but also about our relationship to those lands. How we all think about, interact with, and ultimately treat those lands must be a reflection of how we interact with and treat ourselves, individually and collectively. It is true that sometimes we may not treat ourselves very well, abusing our minds, bodies, and souls. It is a definite human trait to deviate from a path of harmony and balance. I know my deviations exist, and they have left scars upon my inner soul that remind me of the disconnect I can create.

However, out there, in the wild natural world, I believe is our opportunity to know, see, and feel that there is this concept of balance. Nature is neither compassionate nor harsh; it does not judge nor condemn; it is simply there, as it has been since creation. I believe it is we who seek to humanize it, to bring it down to our level, so that we can attempt to put some meaning into it. Perhaps it should be the other way around; that we should attempt to find the meaning inherent in Nature, to be a part of Nature. Thus in the end, we too can achieve balance in our lives, taking and giving, learning and teaching, and ultimately living contently and dying fulfilled.

Notes

1. Leigh J. Kuwanwisiwma, "Hopi Navotiat, Hopi Knowledge of History: Hopi Presence on Black Mesa," in *Prehistoric Cultural Change on the Colorado Plateau: Ten Thousand Years on Black Mesa*, eds. Shirley Powell and Francis B. Smiley (Tucson: University of Arizona Press, 2002), 161.
2. Leigh J. Kuwanwisiwma and T. J. Ferguson, "Ang Kuktota: Hopi Ancestral Sites and Cultural Landscapes," *Expedition* 46, no. 2 (2004): 25–29.
3. Leigh J. Kuwanwisiwma, "Hopi Understandings of the Past: A Collaborative Approach," in *Public Benefits of Archaeology*, ed. Barbara J. Little (Gainesville: University Press of Florida, 2002), 46–50; Kuwanwisiwma, "Hopi Navotiat, Hopi Knowledge of History," 161–163.
4. David Petersen, *Heartsblood: Hunting, Spirituality, and Wildness in America* (Washington, D.C.: Island Press, 2000), 112.
5. T. J. Ferguson, "Öngtupqa niq Pisisvayu (Salt Canyon and the Colorado River): The Hopi People and the Grand Canyon," final ethnohistoric report for the Hopi Glen Canyon Environmental Studies Project, produced by the Hopi Cultural Preservation Office, 1998.
6. Micah Lomaomvaya, T. J. Ferguson, and Michael Yeatts, "Öngtuvqava Sakwtala: Hopi Ethnobotany in the Grand Canyon," produced by the Hopi Cultural Preservation Office, 2001.
7. Petersen, *Heartsblood*, 96.
8. Excerpt from a speech Vernon gave at the 1991 Hopi Cultural Preservation Day. Transcript in possession of author.

Sonia Dickey, photo by Poetic Images by Deanna

Sonia Dickey

SONIA DICKEY GRADUATED IN MAY 2011 WITH HER PHD IN HISTORY. Her dissertation, "Sacrilege in Dinétah: Native Encounters with Glen Canyon Dam," considers the Navajos' role in developing the Colorado River. She has worked for the last ten years in the writing and publishing industry. Sonia currently lives in Gainesville, Florida, where she is the rights and permissions manager/grants coordinator at the University Press of Florida.

"Don't You Let That Deal Go Down"

Navajo Water Rights and the Black Mesa Struggle

ON SATURDAY, APRIL 17, 1971, A SMALL GROUP OF PROTESTERS GATHERED on Black Mesa, the heart center of both the Navajo and Hopi nations. That morning, chaos and confusion stymied planned demonstrations by Navajos, Hopis, and environmental and student activists. According to *Dine Baa-Hani*, an alternative Navajo newspaper headquartered at Crownpoint, New Mexico, Navajo police and Peabody Coal officials had stolen roadside signs directing dissenters to the protest's original rallying point.[1] By two o'clock in the afternoon, however, participants had reconvened. This time, they formed a fifteen-car convoy led by two motorcycles carrying white flags inscribed "Mother Earth." To prevent demonstrators from entering the company's warehouse district and central office, Peabody Coal constructed a practically defunct roadblock manned by workmen, Navajo police, and two pickup trucks.[2] Protesters infiltrated the blockade, forcing Peabody to abandon its efforts. Company officials then shut down Peabody's warehouse complex and main offices before the close of business.

Despite forty-mile-per-hour spring winds and obnoxious workers "sneer[ing] and [throwing] fingers at Indian people," demonstrators met at company headquarters, where they conducted prayers and delivered speeches in opposition to Peabody's strip-mining operations on Black Mesa. Eighty-three-year-old Asa Bazhonoodah, a Navajo elder and herder who had lived on Black Mesa all her life, told the crowd, "This is not legal. We the people of Black Mesa have not surrendered . . . We have fought Kit Carson and weren't marched to Fort Sumner. This is us, the Navaho [*sic*]. We will never surrender!"[3] Hopi traditionalist David Monongye took the stage after the old woman. In spite of Peabody employees heckling him, old man Monongye prayed and offered corn pollen. Hopi interpreter Thomas Banyaca, born in 1909 to the Coyote clan, then proclaimed, "We shall unite and fight . . . not through war, machines, money, business, or violence shall we fight, but through Spiritual Power, will we fight."[4] The small assembly cheered and then challenged Peabody officials "to say their piece." When Peabody's upper echelon refused, demonstrators piled into their cars and headed home.

Black Mesa's rufous-hued plateau gently rises 8,110 feet from the Little Colorado River northeast to its northern rim at Kayenta, Arizona. A distinct elevated landmass spanning four thousand square miles, Black Mesa sits squarely in the center of the Colorado Plateau. It encompasses parts of the Navajo Reservation as well as Hopiland's First, Second, and Third Mesas, each home to villages built around sacred springs fed by the Navajo Aquifer.[5] Once a Pleistocene lake, Black Mesa now resembles a human hand tracing the contours of this ancient body of water. "Where the fingernails are," Bazhonoodah's grandson Key Shelton claims, "that's the Hopi mesas."[6]

For thousands of years, the lake embraced vibrant forests and plant life before decaying into a bog that, over time, hardened to twenty-one billion tons of high-grade, low-sulphur coal, the female mountain's lungs according to Diné mythology and one of the largest coal deposits in the United States.[7] Situating Black Mesa within the Diné worldview and connecting one landform to another, Shelton recites his grandmother's geomythic mapping of the natural ziggurat:

> Navajo Mountain is the head of "she" mountain, and Big Mountain, the most prominent on Black Mesa, is the liver to that female mountain . . . The head of the male mountain is at Ganado, where the rock sits and there is a stand of trees that look as though they are tied together. Then, on the road to Page is a tall, narrow rock—that rock is a child, and where the road turns in a long, narrow curve, it is Mother Earth's arm holding her grandchild. The waterways connect the mountains. They run in a circle, clockwise from east to west . . . all these rivers form the boundaries of our sacred universe. The rivers are intertwined with each other, and like a young couple they flow—laughing and twisting and turning—linking the mountains into a communication system that is the Navajo Way.[8]

In the mid-1960s, when Peabody Coal obtained mining leases from both the Navajo and Hopi tribal councils, Black Mesa's dry washes shimmered with shiny stratums of bituminous coal. Although the company claimed that shale and rock covered most of the deposits "up to 120 feet deep," seams three to eighteen inches thick reached the surface, allowing Peabody to employ both surface- and strip-mining techniques in its quest for the raw material.[9] Peabody's relentless razing of Black Mesa in search of coal severely eroded the region's soil, reducing its fertility; polluted the area's rivers and streams; scarred the surrounding landscape; destroyed native plants and animals; and resulted in the forced removal of thousands of Navajos from their traditional homeland. Enormous draglines and bulldozers raped and pillaged the land, terrorizing its Native inhabitants whose families had lived there for generations. Indeed, aural historian Jack Loeffler, cofounder of the Black Mesa Defense Fund, witnessed Peabody Coal bulldoze an elderly Navajo woman's hogan. "The woman's world of a lifetime," he writes, "disappeared before her weeping eyes in a cloud of dust, and then she was homeless."[10]

Navajos and Hopis alike expressed their despondency over Peabody's destruction of their homeland. "Now there is a big strip mine where coal comes out of the Earth," John Lansa told other Hopi traditionalists gathered in Hotevilla, Arizona. "This makes upheaval on the land. They cut across our sacred shrines . . . Peabody is . . . destroying the sacred mountain. What they take away from our land is being turned into power to create even more evil things."[11] Recalling a recent trip to gather medicine, Bazhonoodah also lamented the effects of coal mining on Black Mesa. "I have gone three times to look for herbs," the old woman told members of the U.S. Senate Committee on Interior and Insular Affairs. "I couldn't recognize the place where we find them . . . I couldn't find my way around the mountain because it was so distorted."[12] Paul A. Begay, a twenty-year veteran of the Navajo Tribal Council who championed the value of homeland, remembered his grandfather Dághii Sanii (Old Mustache) "tell[ing] me that below one foot of topsoil lie important minerals that promote growth of all vegetation . . . And Black Mesa has in it all the valued possessions[:] turquoise, sheep, horses, cattle, and all that is important to our lives." Peabody Coal "has moved in, started digging into the Earth's depths, and is destroying our Navajo religious values and wealth. . . . [The strip-mining,]" Begay concluded, "just must be stopped."[13]

In the mid-sixties, the Diné still lived in isolated family groups in hogans scattered along the backside of Dzilíjiin, or Black Mountain, also known as Black Mesa, where they herded sheep, goats, and horses. Likewise, traditional Hopis inhabited the escarpment's southern promontories, raising corn, beans, and squash just as their ancestors had done for generations. The Navajos' and Hopis' protests against Peabody Coal Company, the nation's largest producer of coal with mining operations in ten states and, at the time, a subsidiary of the Kennecott Copper Corporation, echoed sentiments of Red Power, intertribalism, and indigenous sovereignty. More important, it exemplified a rising intolerance of colonial powers, which sometimes included their own tribal governments, among traditional Navajos and Hopis. These colonial forces historically sought to destroy Native communities by suppressing their cultures, pillaging their land, and stealing their water. Indigenous peoples have always resisted outside interests, but during the late sixties and early seventies, Native communities all over Indian Country embraced a new doctrine of Red Power, one that emphasized treaty rights, tribal sovereignty, religious freedom, self-determination, and intertribalism, all laced with sentiments of decolonization. "People . . . from across the sea have come and mercilessly slaughtered our people," Bazhoonodah told the audience. "You have won the war," she cried. "Go hame [*sic*] back across the sea. . . . We cannot exist in this concentration camp any longer."[14]

The gut-wrenching story of strip-mining on Black Mesa is intricately tied to the multilayered, interconnected history of indigenous water rights, water reclamation, and energy development in the American Southwest. Plans for various infrastructure projects began in earnest during the waning years of the Progressive Era and culminated with the passage of the Colorado River Storage Project Act of 1956, promoting Glen Canyon Dam as its sensational star, and Congress's approval of the Central Arizona Project (CAP) as part of the Colorado River Basin Project Act of 1968. These federal laws and interstate compacts historically ignored indigenous claims to the region's rivers and streams.

During negotiations for the Central Arizona Project in the mid-1960s, Arizona politicians, spearheaded by the state's senior senator Carl T. Hayden, continued that tradition, snubbing Navajo property rights and ignoring the tribe's historical land and water claims. The tribal council, emboldened by their newly elected chairman Raymond Nakai, who preached tribal sovereignty and Indian nationalism, originally helped block plans for a hydroelectric dam at Marble Canyon in Grand Canyon as part of the CAP but later endorsed the construction of the Navajo Generating Station (NGS), relinquishing control of Diné rights to the Colorado River. Nakai and the tribal council cared less about saving the Grand Canyon than asserting Navajo self-determination and securing economic security for the People through extractive industries, especially coal mining on Black Mesa. Arizona's political players infuriated Nakai and members of the Navajo Tribal Council when they openly ignored Diné claims to Marble Canyon and failed to include tribal leaders in discussions for the CAP. Navajo officials shut down further negotiations with CAP proponents once they realized that Hayden, Senator Barry M. Goldwater, and Arizona representatives John J. Rhodes and Morris K. "Mo" Udall intended to filch Navajo land to fuel the massive public works project.

If constructed, the dam at Marble Canyon would have flooded forty-six miles of Diné Bikéyah (the Navajo Reservation). Moreover, early negotiations with the Diné had established tribal rights to waterways, including the Colorado and San Juan rivers, defining reservation boundaries. The treaty signed between the Navajos and the U.S. government in 1868, for example, guaranteed the People a permanent homeland, which implied tribal access to southwestern rivers and streams.[15] In 1908, the U.S. Supreme Court acknowledged the inherent value of treaties between American Indians and the United States when it reserved water rights for Native communities in its landmark decision *Winters v. United States*. Unfortunately, the court's ruling stopped short of quantifying those privileges.

The case centered on the Assiniboines and Gros Ventres living on the Fort Belknap Reservation in north-central Montana. Originally part of a vast area set aside in 1855, but reduced to fourteen hundred acres in 1888, the reservation adjoined the Milk River, a tributary of the Missouri River. In the late 1880s, an influx of Euro-American ranchers and farmers accompanied the arrival of the Great Northern Railroad bisecting Montana. The new settlers immediately established small isolated farming communities all along the Milk River and diverted its waters to their fields, ignoring their indigenous neighbors. Farmers and ranchers fought among each other over access to the Milk. Canada, along with additional upstream users, appropriated a large portion of the river prior to its waters flowing downstream. Most summers, the Milk would dry up completely. By 1900, this reality had forced the nascent Reclamation Service, a forerunner to the U.S. Bureau of Reclamation, to consider the feasibility of diverting water from the St. Mary River, which effuses from Gunsight Mountain in Glacier National Park, through a gravity canal to the Milk River. In 1904 and 1905, a severe drought gripped the entire region. This climatic catastrophe, combined with upstream diversions, drained the Milk River dry before it ever reached Fort Belknap. "Our meadows are now parching up," Superintendant William R. Logan wrote to his superiors at the Office of Indian Affairs. "The Indians have planted large crops and a great deal of grain. All this will be lost unless . . . radical action is taken . . . to make the settlers above the Reservation respect our rights."[16]

Given the federal government's obsession with assimilating the nation's indigenous populations by turning Indians into sedentary farmers, Logan's call did not go unanswered. U.S. Justice Department attorney Carl Rasch quickly requested a court order to restrict Henry Winter (a clerical error resulted in his name being spelled *Winters* in the Supreme Courtcase) and other non-Native farmers and ranchers from diverting water needed by Indian inhabitants on the reservation. Rasch argued that Fort Belknap held riparian rights to the Milk River. The Riparian Doctrine, the traditional water law governing the eastern United States, asserts that everyone living along a river shares in its bounty.[17] Realizing his argument presented numerous pitfalls when applied to western water, Rasch also maintained that the treaty between the United States and the Assiniboines and Gros

Ventres created a water right. Moreover, he claimed no state could repudiate the federal government's privileges as a landowner and trustee for Native peoples.[18] Much to the settlers' chagrin, a federal district court upheld the Indians' reserved rights to water and invented a wholly new water doctrine based on treaty promises. "When the Indians made the treaty granting rights to the United States," Judge William H. Hunt wrote, "they reserved the right to the use of the waters of the Milk River."[19]

Not surprisingly, Winter and the other appellants appealed, igniting a firestorm of controversy that lasted three years and spread to the U.S. Supreme Court.[20] In January 1908, the Supreme Court ruled in favor of the Assiniboines and Gros Ventres. The court's eight-to-one decision reflected the nation's commitment to assimilating Native peoples into the dominant culture through farming, not its concern for indigenous ways of life. It concluded that the government, through its treaty with the Assiniboines and Gros Ventres, reserved the waters of the Milk River "for a use which would be necessarily continued through the years."[21] In other words, when Congress set aside land for an Indian reservation, it also retained the water rights to fulfill the reservation's purpose. Although this case, the first in a long sequence of litigations known collectively as the Winters Doctrine, secured Native claims to water, it created a great deal of uncertainty surrounding indigenous water rights. The ruling's ambiguity allowed the federal Bureau of Reclamation to launch a well-organized resistance against the mandate, an all-out assault on Native peoples that continued for nearly a century.[22]

Forty years after the implementation of the Winters Doctrine, the Upper Colorado River Basin states of New Mexico, Colorado, Utah, Wyoming, and Arizona ratified the Upper Colorado River Basin Compact of 1948. Although the agreement ignored the needs of Native peoples dependent on Colorado River water, it apportioned the 7.5 million acre-feet of Upper Basin water, outlined in the Colorado River Compact of 1922, among the individual Upper Basin states. More important, it guaranteed Arizona, which rested on each side of the dividing line at Lee's Ferry, 50,000 acre-feet of river water annually. Given that almost all of Arizona's lands in the Upper Basin lie in the geopolitical boundaries of the Navajo Reservation, the Upper Colorado River Basin Compact indirectly granted the Diné a significant portion of the Colorado River, an allotment coveted by Arizona's midcentury politicians who pushed for the creation of a central Arizona project.[23]

These figures, however, failed to represent an accurate amount of water flushing the Colorado River system each year. The Colorado River Compact of 1922 simply neglected to observe environmental conditions that often produced an erratic source of water, and it certainly adhered more to arbitrary geopolitical boundaries separating one state, or basin, from another than to natural borders defined by watersheds. Folklorist Hal Cannon denotes a vast difference between wet and dry years in an arid ecosystem. "The drama between those two amounts are so extreme," Cannon contends, "that it is hard to believe that you can assume that it is always going to be a constant to just turn on your tap, because Nature don't work that way." Colorado River historian Norris Hundley Jr. flat-out insists, "The river flow assumptions were in error." When congressional debates over the construction of the Boulder Canyon Project, which included Boulder Dam (now Hoover Dam), erupted in the 1920s, the Colorado River Board of California realized that the average annual flow of the river's main stem hovered somewhere around 14 million acre-feet of water not 17 or 18 million acre-feet as the compact of 1922 predicted.[24]

The Colorado River Compact's miscalculations of realistic water flow at once undermined the entire premise of southwestern water law. In addition, the compact failed to clarify rights to surplus river water and never allocated specific amounts to states within each basin. Instead, it randomly split the Colorado River between the Upper and Lower Basins and awarded each an absolute sum of flow. Nevertheless, the Río Colorado presented possibilities for economic development unmatched by any other water source in the arid Southwest, and farmers, civic boosters, politicians, engineers, businessmen, and government officials all wanting to control the region's most valuable natural resource have fought a myriad of battles over its destiny.[25] The compact's ambiguities, for example, fanned intense conflicts already underway between the Grand Canyon State and its western neighbor. Indeed, clashes between Arizona and California over a reliable supply of water, one large enough to sustain growth, began at the turn of the twentieth century before Arizona even became a state.

According to historian Marc Reisner, the controversy stemmed from the Salt River Project, a series of dams erected on the Gila River in mountain canyons east of Phoenix during the early twentieth century.[26] The Gila and its primary tributaries, the Salt and Verde rivers, serves as Arizona's only "indigenous river of consequence." Flowing from the Black Range of the Gila National Forest in New Mexico along the western slopes of the Continental Divide, the Gila River drops 650 miles before reaching the Colorado River at Yuma, Arizona. The Gila historically evaporated as it rambled across the scorching plains of the Sonoran Desert, resulting in a fickle water source that necessitated dams as a means to increase storage and decrease water loss. Theodore Roosevelt Dam, as well as others constructed on both the Salt and Verde rivers, provided Arizona 2.3 million acre-feet of water to use for agricultural and municipal purposes.

During negotiations to divvy up the Colorado River between its Upper and Lower Basins, California argued that Arizona already possessed its share of river water owing to the Salt River Project. Arizona maintained that it deserved more than 500,000 acre-feet of water from the main stem of the Colorado River in order to sustain growth. Moreover, Arizona's political, agricultural, and economic interests had long coveted an aqueduct to carry water from the Colorado River to the central part of the state. Mostly local- and state-level endeavors, proposed schemes included a "Highline Canal" 500 miles long, with 60 miles of tunnels and numerous concrete-lined canals designed to transport boats and barges.[27] At the meeting in 1922 at Bishop's Lodge in Santa Fe, New Mexico, however, California vowed to squash Arizona's dream of a federally funded aqueduct. Worried that California would eventually consume Arizona's meager share of the Colorado River; incensed by the compact's inclusion of Colorado River tributaries, particularly the Gila River, in its allocation scheme; and bitter over the possibility of using Arizona's tributaries, especially the Gila, to alleviate "the Mexican burden"; representatives for the Grand Canyon State walked away from Santa Fe without signing the Colorado River Compact of 1922.[28] Arizona eventually acquiesced and ratified the agreement in the mid-1940s, when the United States entered treaty negotiations with Mexico in an effort to secure Mexican claims to the Colorado River. Arguments between the Grand Canyon State and its western neighbor over a central Arizona project continued well into the twentieth century and culminated with an eleven-year court battle that ended on June 3, 1963, when the U.S. Supreme Court finally awarded Arizona its annual apportionment of 2.8 million acre-feet of Colorado River water in its decision *Arizona v. California*. In addition, the judgment divided surplus flows between the Grand Canyon and Golden States fifty-fifty.[29]

The case proved somewhat beneficial to Native communities as well. Similar to the *Winters* decision, *Arizona v. California* maintained that the establishment of a reservation necessarily implied rights to water sources within or abutting reservation boundaries. It also secured enough water for indigenous peoples "to irrigate all the practically irrigable acreage," or PIA, on reservations.[30] Although *Arizona v. California* quantified Indian water based on a reservation's PIA, the court's decision fell short of making the standard a universal and timeless test. This uncertainty eventually forced tribes, including the Navajos, to bargain for their apparent water rights in exchange for economic security, a move that potentially threatened the political and cultural integrity of Native communities.

Although Arizona's powerful water interests had proposed a public works project funded by the state since the early twentieth century, Arizona's politicians wasted no time in their push for a federally subsidized central Arizona project. On June 4, 1963, one day after the U.S. Supreme Court ruled in favor of Arizona receiving its fair share of the Colorado River, Arizona's senators Hayden and Goldwater introduced to the Senate Subcommittee on Irrigation and Reclamation a bill authorizing a central Arizona project. Hayden and Goldwater's proposal called for a system of conduits, canals, and pumping plants to divert water from Lake Havasu, impounded behind Parker Dam on the Arizona-California border, via the Havasu Intake Channel Dike to Orme Dam at the confluence of the Salt and Verde rivers on the Yavapais' Fort McDowell Reservation near Phoenix.[31] In addition to a main canal carrying water from Lake Havasu, the plan included regulating facilities; a hydroelectric dam, reservoir, and power plant at Bridge Canyon on the Hualapai Reservation near Peach Springs, Arizona; electrical transmission lines; Buttes and Hooker Dams on the Gila River to provide conservation storage, flood and sediment control, and recreation opportunities; Charleston Dam and its subsequent reservoir on the San Pedro River southeast of Tucson; and the Tucson and Salt-Gila aqueducts.[32] Likewise, in January 1964, Secretary of the Interior Stewart L. Udall published the U.S. Bureau of Reclamation's *Pacific Southwest Water Plan*. He proposed to redirect water from the verdant Pacific Northwest to the arid Southwest through a series of tunnels, ducts, and canals.[33] The increased water flow through this interbasin transfer would aid in new dams and diversions designed to remedy the growing Southwest's water shortage. Calling for two phases of construction over thirty-five years and costing more than four billion dollars, the secretary's blueprints centered on Arizona's long-awaited aqueduct system. To power the delivery of water from the Colorado River to farmers and cities in the state's central valleys, Secretary Udall, backed by Bureau of Reclamation commissioner Floyd E. Dominy, suggested building two hydroelectric dams on opposite ends of the Grand Canyon, including one at Bridge Canyon and the other at Marble Canyon on Navajo land.

"[I]n 1949, when I was on the Sierra Club Board of Directors," David R. Brower once recalled, "I voted for both dams [in Grand Canyon]."[34] Evidently relying on a faulty memory, he later admitted he first became aware of plans for the Central Arizona Project during the early 1950s. At the time, the conservation movement battled the U.S. Bureau of Reclamation to save Dinosaur National Monument from Echo Park Dam and its promise of ecological destruction as dictated in previous versions of congressional bills containing legislation for the Colorado River Storage Project. The completion of Glen Canyon Dam in January 1963, and its subsequent reservoir, paved the way for major power projects requiring substantial amounts of water, such as the Navajo Generating Station built on the shores of Lake Powell during the early 1970s. Indeed, Brower alluded to plans for a gigantic buildup in the American Southwest, a juggernaut of hydroelectric dams, power lines, and coal-fired power plants to light up the region's swelling urban centers, including San Diego, Los Angeles, Las Vegas, Phoenix, Tucson, and Albuquerque. Of course, federal officials and congressional leaders touted agriculture as the primary reason for the erection of dams up and down the Colorado River, but Brower and his conservation cohort "could count on developers getting hold of the water one way or the other." Loeffler validated Brower's initial suspicions, adding, "The Southwest was being raped by developers and mining companies. The land was being pillaged, the waters fouled, the air smoke-dimmed, traditional cultures left bereft. Power lines marched across the land like electric kachinas."[35]

The Southwest experienced tremendous growth immediately following World War II. The Cold War essentially propelled federal expenditures as the Department of Defense consumed the region's open space to support its military bases, naval shipyards, air force landing strips, and nuclear test sites. Defense spending had ignited a firestorm of development rapidly spreading to southwestern urban centers. Phoenix proved the most remarkable. By the mid-1950s, Arizona ranked as the fastest-growing state in the union, and Phoenix stood poised for a "long, explosive boom." Author Edward Abbey referred to the population upsurge plaguing Arizona as "the Blob . . . a mad amoeba . . . growing *growing* ever-GROWING."[36] City planners set "all the machinery in place."[37] In order to attract new settlers, civic boosters, primarily financed by military installations and the aeronautics industry, spearheaded capital improvements, including roads, airports, water supplies, sewer systems, and parks.[38] Yet business leaders and congressmen surmised that Phoenix still lacked capital and water. "Arizona is at a crisis point," Hayden testified at the CAP hearings in 1963. The state "urgently needs more water . . . to meet her rapidly expanding domestic requirements."[39] Indeed, the Grand Canyon State demanded water and electricity to illuminate its developing cities. Fortunately, Phoenix rested on the edge of the Colorado

Plateau, a rich storehouse of natural resources and home to the nation's largest Native community, both crucial for a colonialist system dependent on Indian assets to feed the dominant culture.

From 1963 to 1968, Arizona's political players, including Secretary Stewart Udall; Senators Hayden and Goldwater; and Representatives Rhodes and Mo Udall, Stewart's younger brother and successor in the House, fought Sierra Club executive director Brower and other conservation leaders over the inclusion of dams in the Grand Canyon as part of the CAP.[40] Somewhere between saving Dinosaur National Monument and abandoning Glen Canyon, Brower seemingly learned a lesson on the value of a wild river. Brower confessed that his "big slip" occurred "when I was at least momentarily for . . . a taller Glen Canyon Dam. . . . That's a story I'd rather forget," he admitted. Although the Colorado River Compact of 1922 necessitated a dam at or near the mouth of Glen Canyon owing to its close proximity to Lee's Ferry, Brower still regretted his lack of concern for the labyrinthine abyss. "That's where I began to learn," Brower divulged, "that you don't give anything away that you haven't looked at."[41] Despite Brower's blunder, the U.S. Bureau of Reclamation had always intended to construct a dam at Glen Canyon near the compact's dividing line at Lee's Ferry. Brower's epiphany, however, perhaps motivated his all-out assault on the CAP's inclusion of dams in the Grand Canyon. Brower used his clout within the budding environmental movement as well as Sierra Club monies to launch an impressive publicity campaign. Full-page ads appeared in national newspapers, including the *New York Times*, galvanizing the American public into action. For its part, Arizona, in cahoots with the Bureau of Reclamation, continued to fight for a federally funded central Arizona project, complete with dams on both ends of the Grand Canyon. Stewart Udall's post as secretary of the Interior no doubt bolstered the Grand Canyon State's position within President Lyndon B. Johnson's administration.

Arizona's Native communities played an important part, too. Historian Byron Pearson argues that both environmentalists and CAP enthusiasts capitalized on increasing public concern about civil rights and race relations dominating mainstream America during the mid-1960s. According to Pearson, both sides wooed the Hualapai and Navajo nations in an attempt to obtain indigenous support for their respective agendas. Ultimately, Pearson asserts, they wanted to influence public opinion on debates surrounding the Central Arizona Project.[42] Although they promised the Hualapais tremendous economic benefits from Bridge Canyon Dam, proponents of the Grand Canyon dams never offered the Navajo Nation compensation for a structure at Marble Canyon. Goldwater told geographer Stephen C. Jett, then a young professor at the University of California, Davis, that as far as he knew, the Navajos "possessed no legal rights" to Marble Canyon and "had not been consulted." The senator admitted that the Department of the Interior refused to acknowledge Diné claims to the site.[43]

The failure of CAP devotees to include the country's largest Indian group as beneficiaries arguably hindered the advancement of their dam agenda. To make matters worse, CAP lobbyists had agreed to pay the Hualapais $16 million for the construction of Bridge Canyon Dam on their land near Peach Springs, Arizona, while denying Navajo claims to the Marble Canyon site.[44] Their actions no doubt infuriated Navajo tribal chairman Raymond Nakai and other Diné leaders and convinced them to reconsider their support for hydroelectric structures in Grand Canyon. The Navajo Tribal Council backed proposals calling for a dam at Marble Canyon as late as 1961, when Arizona still planned on constructing a state-funded central Arizona project. On May 3, 1967, however, Nakai wrote New Mexico senator Clinton P. Anderson, chairman of the Senate Subcommittee on Water and Power Resources, outlining the tribal council's new stance. In his letter, Chairman Nakai expressed frustration with both the Arizona Power Authority and the Federal Power Commission, state and federal entities charged at different times with constructing and maintaining a dam at Marble Canyon, and the agencies' refusal to offer the Navajos "reasonable compensation" for the use of their land. The chairman then dropped a bombshell. Nine months earlier, on August 3, 1966, the Navajo Tribal Council, asserting Diné rights as a sovereign nation, had passed resolution CAU-97-66. In its twenty-nine-to-two decision, the council cited the "ruthless character" of CAP lobbyists, including Mo Udall and his brother, and revoked the tribe's initial support for a dam at Marble Canyon.[45] The declaration directly addressed the Udall brothers, maintaining that "you have violated the policy of the administration, you have violated the wishes of the President, you have violated the Park Service, . . . [and you] have ignored the property rights and interest of the Navajo Tribe."[46]

Despite CAP enthusiasts' repudiation of Diné claims to the upper end of Marble Canyon, resolution CAU-97-66 insisted that the proposed dam would flood forty-six miles of scenic reservation land. "Our resolution makes quite clear," tribal councilwoman Annie Dodge Wauneka contended, "that the great expenditures contemplated for the two dams . . . are wholly unnecessary." Instead, the Navajo Tribal Council wanted to exploit the "huge deposits of coal" buried in the Four Corners and on Black Mesa.

"We have learned," Wauneka told the press, "that hydropower cannot possibly compete with cheaper thermo-electric plants stoked with Reservation coal," and "we have concluded agreements with Peabody Coal Co. and Utah Construction Co. for burning Navajo coal to power what are probably the largest generating plants in the country."[47]

On February 1, 1964, Peabody Coal Company's predecessor in interest, Sentry Royalty Company, signed a ten-year lease with the Navajo Tribal Council to extract Dzilíjiin's shiny stratums of bituminous coal. According to contract number 14-20-0603-8580, or lease no. 8580 for short, Sentry Royalty agreed to pay the Navajo Nation a variable rate ranging from twenty-five cents to thirty-seven and a half cents per ton of coal sold and utilized off the reservation in exchange for the right to mine approximately 25,000 acres of Navajo land on Black Mesa near Kayenta, Arizona. When coal from the Kayenta Mine remained on the reservation, the Diné pocketed between twenty and thirty cents per ton, depending on the total amount sold.[48] In addition, Sentry Royalty rented the Kayenta Mine from the Navajos for one dollar per acre per year for the agreement's first five years. In other words, the Navajo Tribe initially received $125,000 for the use of their land, excluding any royalties from the sale of coal.[49] Finally, Sentry Royalty agreed to hire "Navajo Indians when available for all positions," with one caveat: the company, or lessee, judged the qualifications of potential workers.[50]

At about the same time as the Navajo Tribal Council entered negotiations with Sentry Royalty, Secretary Udall realized that legislation for a central Arizona project must contain alternative sources of power. The Navajos' rejection of Marble Canyon Dam, combined with fierce public outcry over inundating parts of the Grand Canyon, persuaded the secretary to consider other possibilities for fueling the CAP. Although Udall asked his Interior underlings to explore alternate options for energy production, he knew about the mining lease between Sentry Royalty and the Navajo Nation. The Indian Mineral Leasing Act of 1938 required the secretary's approval of all agreements between Native peoples and commercial enterprises in an effort to ensure Indian communities received maximum profits and a lease's terms represented their best interest. In May 1965, commissioner of Indian affairs Philleo Nash sent Secretary Udall a memo informing him that Peabody had proposed to install two coal-fired steam-generating units of one thousand megawatts each and had requested to use approximately two thousand acres of Navajo land close to Lake Powell for the new plant.[51] In addition, Peabody anticipated burning up to six million tons of Black Mesa coal per year and needing forty thousand acre-feet of water annually to operate the generating station.[52] The latter would no doubt come from Arizona's (read: the Navajos') heretofore unused yearly apportionment of fifty thousand acre-feet of Colorado River water allotted to the Grand Canyon State by the Upper Colorado River Basin Compact of 1948. In addition, based on the royalty rates summarized in its agreement with Sentry Royalty, the Navajo Tribal Council would potentially receive at least 1.2 million dollars each year from Dziłíjiin's coalfields.

On June 6, 1966, the Navajo Tribal Council signed a second ten-year mining lease with Sentry Royalty Company. This time, the agreement covered mining operations on 40,000 acres of the Navajo-Hopi Joint Use Area (JUA), 2.5 million acres of land "shared" between the two tribes and the focal point of a hotly contested land dispute that erupted when Peabody Coal Company first discovered coal within JUA boundaries in 1909.[53] The two leases totaled approximately 100 square miles loaded with 337 million tons of coal.[54] The raw material supplied the Mohave Generating Station at Bullhead City, Nevada, for nearly thirty years, and it continues to feed the Navajo Generating Station at Page.

The Department of the Interior, working closely with Peabody Coal, seemingly orchestrated the industrialization of Dinétah (Navajo homeland) in order to proceed with Reclamation's stratagems for Arizona's central valley. On August 17, 1966, just two weeks after the tribal council condemned the proposed Grand Canyon dams and announced their plans for Black Mesa, James H. Krieger of Best, Best, and Krieger, a California-based law firm representing Peabody's interests in the West, informed the Department of the Interior that Peabody's "investor group" would place the Arizona power plant "next on the list" if the agency could guarantee Peabody "early assurance of the water."[55] As a bonus, Peabody estimated that construction would require approximately three hundred workers.

The following February, six months after Peabody's proposition, Udall rescinded his support for hydroelectric dams in the Grand Canyon and lobbied, instead, for a coal-fired power plant on the Navajo Reservation abutting Lake Powell to fuel the CAP. At the time, Udall adulated the benefits of mineral extraction for the Navajo Nation but confessed years later that "the whole Central Arizona Project would have failed . . . [without] an energy source." Since "nuclear power was not sufficiently advanced at the time," Udall maintained, "some of the Arizona people and some of the power companies got busy and put together this Page Power Plant."[56] Given Peabody's close working relationship with Western Energy Supply and Transmission Associates (WEST), a united consortium of twenty-three "investor-owned and state, municipal, and federal" electric utilities, "some of the power companies" Udall referenced no doubt helped devise this new plan.[57] Navajo leaders also realized the potential monetary benefits of extracting raw materials. In September 1966, the *Arizona Republic*, a Phoenix newspaper, announced an agreement authorizing the use of Black Mesa coal for the production of electricity at a coal-fired steam-generating plant near Las Vegas in Clark County, Nevada. The facility, dubbed the Mohave Steam Station, would supposedly yield more than $30 million in royalties for both the Navajo and Hopi nations.[58] Not surprising, then, council members supported Udall's recommendation for a thermal-generating plant near Page in lieu of hydroelectric dams to provide pumping power for the Central Arizona Project.[59] The new plant promised the People jobs and additional income, longtime priorities of the Navajo Tribal Council, and guaranteed the Southwest's urban centers a seemingly endless supply of electrical energy.

Historian Alvin M. Josephy Jr., who actively protested strip-mining on Black Mesa during the early 1970s, accused the U.S. Bureau of Reclamation of facilitating the uninhibited growth of southwestern cities.[60] During congressional hearings for the Colorado River Storage Project, approved in 1956, for example, Reclamation had lauded Glen Canyon Dam and its production of hydroelectric power, while purportedly negotiating with private utility companies for coal-fired power plants. "The bureau had the usual self-serving reasons for busying itself with these added projects," Josephy penned in a damning piece on "the murder of the Southwest" published by *Audubon* magazine in the summer of 1971. "As everyone knows," he wrote, "without projects there is no need for the [agency]."[61] Indeed, Reclamation divvied up, transported, and delivered its "rivers of empire" through whatever means necessary to reclaim the arid Southwest, to "green it up" as historian Patricia Nelson Limerick quips.[62]

In spite of Diné claims to the Colorado River, Interior assured Peabody access to Arizona's "unused" portion of the waterway. On December 11, 1968, the Navajo Tribal Council passed resolution CD-108-68 essentially waiving Navajo rights to Colorado River water in exchange for "limited promises by a consortium of utilities" to light up the reservation's hogans.[63] The council agreed to limit Diné demands on 50,000 acre-feet of water allocated to Arizona each year under the Upper Colorado River Basin Compact of 1948. Instead, council members granted 34,100 acre-feet of the tribe's annual apportionment to the Page power plant in order to water its cooling towers.

The tribal council mandate ostensibly negated Diné claims to the Colorado River in favor of a coal-fired power plant that promised economic security and jobs for the People, as well as electricity. Interestingly, fragmentary evidence suggests Secretary Udall never approved resolution CD-108-68. His failure to endorse the agreement renders it virtually ineffective and does not legally bind the tribe, freeing the Diné to claim more than 50,000 acre-feet of water for tribal lands situated in Arizona's Upper Basin.[64] Today, individual members of the Navajo Nation argue that the Colorado River belongs to the Diné and holds tremendous economic potential for the tribe if their government leaders, often influenced by white lawyers, would only assert the People's rights.[65]

By May 1969, the tribal council had entered negotiations with a conglomerate of utility companies to lease almost two thousand acres of reservation land for the Navajo Generating Station. According to the *Navajo Times*, the Navajo Tribal Utility Authority (NTUA) would share in the plant's electrical output. Bureau of Indian Affairs regional director Graham Holmes told reporters, "The proposal fits in with overall plans to develop the reservation."[66] Holmes's admission highlighted the efforts of both the Department of the Interior and the Navajo Tribal Council to industrialize Dinétah during the postwar era. In this case, their actions displayed utter disregard for the People living in the shadows of Peabody's draglines, those traditional Navajos whose families had inhabited Dziłíjiin for generations. The Salt River Project, responsible for the operation and maintenance of the power plant, built the Navajo Generating Station in three phases. It completed the third and final unit in 1976. With smoke stacks seven stories high, the smoke-billowing monster devours 23,000 tons of coal daily and swallows 270,000 gallons of Navajo water, pumped out of Lake Powell, per minute. The plant delivers 2,310 megawatts of electricity to its customers via 800 miles of transmission lines marching across the desert like storm troopers from *Star Wars*.[67] Interior Secretary Udall, worried that Congress would never approve CAP legislation authorizing hydroelectric dams in the Grand Canyon, worked closely with Peabody and the region's utility companies to develop alternate solutions, eventually concocting plans for the NGS to provide electricity to power the Central Arizona Project.[68] The Bureau of Reclamation, an agency of the Department of the Interior, currently owns 24.3 percent of the electricity effusing from the Page power plant, while the Salt River Project claims 21.7 percent, Nevada Power receives 11.3 percent, the Arizona Public Service gets 14 percent, Tucson Gas and Electric maintains 7.5 percent, and, finally, Los Angeles Water and

Power holds 21.2 percent.[69] The last figure evidently reflects the longtime controversy between Arizona and California over which state receives the most benefits from the Colorado River. "We broke the mold with that plant," Secretary Udall reminisced.[70]

The origins of strip-mining coal on Black Mesa can be found among the pages detailing a multilayered, interconnected history of indigenous water rights, water reclamation, and energy development in the American Southwest. The Colorado River Compact of 1922, which for all intents and purposes neglected Native rights to the Colorado River, necessitated the construction of Glen Canyon Dam to transfer Colorado River water from one basin to the next. Likewise, the concrete plug paved the way for major power schemes, such as the Central Arizona Project, requiring substantial amounts of water to illuminate the region's burgeoning postwar cities. Although these federal laws and interstate compacts historically ignored indigenous claims to the region's rivers and streams, the saga nonetheless entangled Native peoples, including Navajos and Hopis, in a watery web spanning both time and space. Likewise, the expansion of southwestern urban centers rested on the region's Native populations who supplied raw materials and natural resources necessary to sustain growth. As debates surrounding the Central Arizona Project demonstrate, Native leaders seemed *just as responsible* for the environmental degradation of their homeland as federal officials orchestrating the entire deal. Although Nakai advocated tribal sovereignty and Navajo self-determination during his election campaign in the early 1960s, his resolve to industrialize Diné Bikéyah in an effort to generate much-needed revenue on the reservation helped tighten the bond between water reclamation and Navajo colonization. Despite huge strides toward asserting the People's power, Nakai and the Navajo Tribal Council resorted to the same type of money-making schemes characteristic of other Navajo leaders during the postwar era. Their efforts reflected an imposed economic system bent on turning habitat into money, all at the expense of their constituents.

Notes

1. For chapter title origin, see Jerry Garcia and Robert Hunter, "Deal," in *The Complete Annotated Grateful Dead Lyrics*, ed. Alan Trist and David Dodd, annot. David Dodd (New York: Free Press, 2005), 155.
2. "Dine-Hopis Protest," *Dine Baa-Hani* (Crownpoint, NM), April 22, 1971.

3. Asa Bazhonoodah, quoted in "Dine-Hopis Protest."
4. "Dine-Hopis Protest"; and Thomas Banyaca, quoted in "Dine-Hopis Protest."
5. Tamisha Grimes, "The Hopi and the Black Mesa: An Argument for Protection of Sacred Water Sites," (master's thesis, University of Kansas, 2008), 5.
6. Key Shelton, quoted in T. C. McLuhan, *The Way of the Earth: Encounters with Nature in Ancient and Contemporary Thought* (New York: Simon & Schuster, 1994), 412.
7. Judith Nies, "The Black Mesa Syndrome: Indian Lands, Black Gold," *Orion* (Summer 2000), http://www.orionmagazine.org/index.php/articles/article/6025; and Shelton, in McLuhan, *The Way of the Earth*, 412.
8. Shelton, in McLuhan, *The Way of the Earth*, 412.
9. Peabody Coal Company, *Mining Black Mesa*, November 1970, in "Problems of Electrical Power Production in the Southwest," U.S. Senate, *Hearings before the Committee on Interior and Insular Affairs*, 92nd Cong., 1st sess. (Washington, D.C.: GPO, 1971), 1628. Strip-mining serves as a type of surface mining during which the operator scalps or removes the surface vegetation (e.g., trees, bushes, and grasses and forbes); bulldozers or scrapers and loaders remove the topsoil; and dynamite blasts the exposed overburden to reach coal seams, which are then fractured by blasts and loaded in trucks or onto conveyor belts. Finally, in theory, the operator redistributes the topsoil and seeds or revegetates the area. Mark Squillace, *The Strip Mining Handbook: A Coalfield Citizens' Guide to Using the Law to Fight Back Against the Ravages of Strip Mining and Underground Mining* (Washington, D.C.: Environmental Policy Institute and Friends of the Earth, 1990), 19.
10. Jack Loeffler, "Tragedy in Indian Country," in *Headed Upstream: Interviews with Iconoclasts* (Tucson, AZ: Harbinger House, 1989), 90. For a slightly altered version of this story, see Jack Loeffler, *Adventures with Ed: A Portrait of Abbey* (Albuquerque: University of New Mexico Press, 2002), 105.
11. John Lansa, quoted in "The Time of Great Purification: As Spoken by Hopi Elders to Jack Loeffler," in *Myths and Techno-fantasies: The First Publication of the Black Mesa Defense Fund* (Santa Fe, NM: Black Mesa Defense, 1971), 11.
12. Statement of Asa Bazhonoodah, in "Problems of Electrical Power Production in the Southwest," 1551.
13. Statement of Paul A. Begay, in "Problems of Electrical Power Production in the Southwest," 1658.
14. Asa Bazhonoodah, quoted in "Dine-Hopis Protest."
15. "Treaty of 1868," in Peter Iverson, *Diné: A History of the Navajos* (Albuquerque: University of New Mexico Press, 2002), 325–34.
16. William R. Logan to Commissioner of Indian Affairs Francis E. Leupp, June 3, 1905, quoted in Norris Hundley Jr., "The Dark and Bloody Ground of Indian Water Rights: Confusion Elevated to Principle," *Western Historical Quarterly* 9, no. 4 (October 1978): 462–63.
17. Daniel McCool, *Native Waters: Contemporary Indian Water Settlements and the Second Treaty Era* (Tucson: University of Arizona Press, 2002), 11.
18. Ibid.

19. Judge William H. Hunt, quoted in ibid., 12.
20. Hundley, "The Dark and Bloody Ground of Indian Water Rights," 463. For an in-depth study of the Winters Doctrine, see John Shurts, *Indian Reserved Water Rights: The Winters Doctrine in Its Social and Legal Context, 1880s–1930s* (Norman: University of Oklahoma Press, 2003).
21. *Winters v. United States*, 207 U.S. 564 (1908). See also McCool, *Native Waters*, 14.
22. McCool, *Native Waters*, 12–15.
23. "Article 3," in Upper Colorado River Commission, *Upper Colorado River Basin Compact*, 81st Cong., 1st session, January 31, 1949 (Washington, D.C.: GPO, 1949), 6.
24. Hal Cannon, interview by Jack Loeffler, and Norris Hundley Jr., interview by Jack Loeffler, in *Moving Waters: The Colorado River and the West*, program 2, six-part radio series, 2001, produced by Jack Loeffler for the Arizona Humanities Council, transcript in possession of author.
25. David P. Billington and Donald Conrad Jackson, *Big Dams of the New Deal Era: A Confluence of Engineering and Politics* (Norman: University of Oklahoma Press, 2006), 102.
26. Marc Reisner, *Cadillac Desert: The American West and Its Disappearing Waters* (New York: Penguin Books, 1993), Kindle ed., chap. 8.
27. Byron Pearson, "'We Have Almost Forgotten How to Hope': The Hualapai, the Navajo, and the Fight for the Central Arizona Project, 1944–1968," *Western Historical Quarterly* 31, no. 3 (Autumn 2000): 297.
28. *Arizona v. California*, 373 U.S. 546 (1963).
29. For a history of California's rivalry with Arizona over Colorado River water, see Rich Johnson, *Central Arizona Project, 1918–1968* (Tucson: University of Arizona Press, 1977), 87–124; and Norris Hundley Jr., *The Great Thirst: Californians and Water: A History*, rev. ed. (Berkeley: University of California Press, 2001).
30. *Arizona v. California*, 373 U.S. 546 (1963).
31. Final legislation for the Central Arizona Project, approved by Congress in 1968, included the Orme Dam, which would have flooded half of the Fort McDowell Reservation. The Yavapais fought the project for ten years and eventually blocked its construction. For a brief history of this water reclamation battle, see Daniel Kraker, "Tribe Defeated a Dam and Won Back Its Water," *High Country News*, March 15, 2004, http://www.hcn.org/issues/270/14627.
32. Carl T. Hayden and Barry M. Goldwater, U.S. Senate, "Central Arizona Project," part 1, *Hearings before the Subcommittee on Irrigation and Reclamation of the Committee on Interior and Insular Affairs on S. 1658*, 88th Cong., 1st sess. (Washington, D.C.: GPO, 1963), 2.
33. Stewart L. Udall, *Pacific Southwest Plan: Report, January 1964* (Washington, D.C.: Bureau of Reclamation, 1964). See also Robert Dean, "'Dam Building Still Had Some Magic Then': Stewart Udall, the Central Arizona Project, and the Evolution of the Pacific Southwest Water Plan, 1963–1968," *Pacific Historical Review* 66, no. 1 (February 1997): 81–98; and Roderick Frazier Nash, *Wilderness and the American Mind*, 4th ed. (New Haven, CT: Yale University Press, 2001), 228.
34. David R. Brower, interview by Jack Loeffler, 1986, transcript in author's possession.

35. David R. Brower, interview by Jack Loeffler, 1986, transcript in author's possession; and Loeffler, *Adventures with Ed*, 104.
36. Edward Abbey, "The BLOB Comes to Arizona," in *The Journey Home: Some Words in Defense of the American West* (New York: E. P. Dutton, 1977), 148.
37. Charles Wilkinson, *Fire on the Plateau: Conflict and Endurance in the American Southwest* (Washington, D.C.: Island Press, 1999), 182.
38. Ibid.
39. Statement of Arizona senator Carl T. Hayden, in U.S. Senate, "Central Arizona Project," part 1, *Hearings before the Subcommittee on Irrigation and Reclamation of the Committee on Interior and Insular Affairs on S. 1658*, 88th Cong., 1st sess. (Washington, D.C.: GPO, 1963), 4.
40. For a detailed history of this battle, see Byron Pearson, *Still the Wild River Runs: Congress, the Sierra Club, and the Fight to Save Grand Canyon* (Tucson: University of Arizona Press, 2002).
41. David R. Brower, interview by Jack Loeffler, 1986, transcript in author's possession.
42. Pearson, "'We Have Almost Forgotten How to Hope,'" 299.
43. Ibid., 308, 310.
44. "Actions of Udalls Rapped by Council," *Gallup Independent* (NM), August 4, 1966.
45. Navajo Tribal Council, "Resolution CAU-97-66," in "Tribe Opposes Dams in Grand Canyon," *Window Rock Navajo Times* (AZ), August 11, 1966. See also Raymond Nakai to Clinton P. Anderson, May 3, 1967, Window Rock, Arizona, in U.S. Senate, "Central Arizona Project," *Hearings before the Subcommittee on Water and Power Resources on S. 1004, S. 1013, S. 861, S. 1242, and S. 1409*, 90th Cong., 1st sess. (Washington, D.C.: GPO, 1967), 714.
46. Ibid.
47. Annie Dodge Wauneka, quoted in "Actions of Udalls Rapped by Council," *Gallup Independent* (NM), August 4, 1966.
48. "Mining Lease, Contract No. 14-20-0603-8580, between Sentry Royalty Company and the Navajo Tribe," in Joint Appendix, vol. 2, *United States of America v. Navajo Nation*, May 13, 2008, http://sct.narf.org/documents/usvnavajonation/joint_appendix_vol_ii.pdf, 191.
49. Ibid., 192.
50. Ibid., 200.
51. Commissioner of Indian Affairs Philleo Nash to Secretary of the Interior Stewart L. Udall, May 12, 1965, Washington, D.C., folder 14, box 154, Stewart L. Udall Papers, Special Collections, Main Library, University of Arizona, Tucson [hereafter SLUP].
52. Commissioner of Indian Affairs Philleo Nash to Secretary of the Interior Stewart L. Udall, August 26, 1965, Washington, D.C., folder 14, box 154, SLUP.
53. U.S. Department of the Interior, Bureau of Indian Affairs, "Mining Lease, Contract No. 14-20-0603-9910, between Sentry Royalty Company and the Navajo Tribe," June 6, 1966, in Joint Appendix, vol. 2, *United States of America v. Navajo Nation*, May 13, 2008, http://sct.narf.org/documents/usvnavajonation/joint_appendix_vol_ii.pdf, 191.

    The Navajo-Hopi land dispute, for all practical purposes, began in 1882, when President Chester A. Arthur established the "Navajo-Hopi Joint Use Area." Owing to the ambiguous nature of Indian law, which tribe held exclusive mineral rights to the sulfur-rich coalfields on the reservation and which tribe had the authority to lease to private mining interests remained unclear. Following the passage of the Indian Reorganization Act in 1934, which overlooked traditional forms of government among Native peoples and, instead, reflected representative democracy run amuck, the Hopis lobbied for the eviction of Navajos from the original 1882 reservation. Although federal officials initially ignored the request, commissioner of Indian affairs John Collier proposed District Six, which included 499,248 acres of exclusive Hopi grazing land centered on the First, Second, and Third Mesas. Despite Hopi refusal to participate in land negotiations with government officials and Navajo representatives, Collier and the Bureau of Indian Affairs implemented District Six in 1936. Eventually established as the Hopis' "official" reservation in 1943, District Six remained at the forefront of Navajo-Hopi affairs. The *Healing v. Jones* decision of 1962 judged that the Navajos and Hopis possess joint and equal rights to the 1882 reservation outside District Six. Arizona congressmen, including Sam Steiger, lobbied to partition this land twelve years later, following the coal-mining leases signed between the Navajos and Peabody Coal Company and the Hopis and Peabody. The U.S. government disregarded arguments from a grassroots alliance between Hopis and Navajos for a traditional settlement of the contested area. On December 22, 1974, Congress passed the Navajo-Hopi Land Settlement Act (a.k.a. Public Law 93-531) in one last attempt to resolve the dispute "legally." The bill ordered representatives from the two tribes to begin a six-month period of negotiations aiming for a settlement. If the tribes did not reach an agreement after six months, the federal government would be authorized to divide the Joint Use Area (JUA) into equal Navajo and Hopi lands. More important, Public Law 93-531 established the Navajo-Hopi Relocation Commission, responsible for drafting a relocation plan for members of one tribe living on lands partitioned to the other. Public Law 93-531 resulted in the relocation of several thousand Navajos from their traditional homeland. The mass removal of Navajos from the JUA represents the largest forced removal of Native peoples since the late nineteenth century. For Navajo perspectives on the Navajo-Hopi land dispute, see John Redhouse, *Geopolitics of the Navajo-Hopi Land Dispute* (Albuquerque, NM: Redhouse/Wright Productions, 1985); and "Monster in Dinétah," in Loeffler, *Headed Upstream*, 101–11. For a detailed history of the Navajo-Hopi Joint Use Area , see Jerry Kammer, *The Second Longest Walk: The Navajo-Hopi Land Dispute* (Albuquerque: University of New Mexico Press, 1980).

54. Alvin M. Josephy Jr., "The Murder of the Southwest," *Audubon*, July 1971, 57.
55. Assistant Commissioner of Indian Affairs E. Reeseman "Si" Fryer to Assistant Secretary Orren Beaty Jr., August 17, 1966, Washington, D.C., folder 14, box 154, SLUP.
56. Stewart L. Udall, interview by Jack Loeffler, n.d., in *Headed Upstream*, 122, 123.
57. For Peabody Coal Company's relationship with Western Energy Supply and

Transmission Associates (WEST), see Wilkinson, *Fire on the Plateau*, 303. For the description of WEST, see Josephy, "The Murder of the Southwest," 54.

58. "Riches for Navajos, Hopis," *Phoenix Arizona Republic*, September 29, 1966.
59. Raymond Nakai to Clinton P. Anderson, May 3, 1967, Window Rock, Arizona, in U.S. Senate, "Central Arizona Project," *Hearings before the Subcommittee on Water and Power Resources on S. 1004, S. 1013, S. 861, S. 1242, and S. 1409*, 90th Cong., 1st sess. (Washington, D.C.: GPO, 1967), 710–11.
60. Josephy, "The Murder of the Southwest," 59.
61. Ibid.
62. See Donald Worster, *Rivers of Empire: Water, Aridity, and the Growth of the American West* (New York: Oxford University Press, 1985). Patricia Nelson Limerick, interview by Jack Loeffler, in *Moving Waters: The Colorado River and the West*, program 1, six-part radio series, 2001, produced by Jack Loeffler for the Arizona Humanities Council, transcript in possession of author.
63. Navajo Tribal Council, Resolution CD-108-68, excerpt in William Douglas Back and Jeffrey S. Taylor, "Navajo Water Rights: Pulling the Plug on the Colorado River?" *Natural Resources Journal* 20 (January 1980): 88.
64. William Douglas Back and Jeffrey S. Taylor, "Navajo Water Rights: Pulling the Plug on the Colorado River?" *Natural Resources Journal* 20, no. 1 (January 1980): 88.
65. For this argument, see Max Goldtooth, Peter McDonald, and Ron Milford, "Navajo Water Rights: Truths and Betrayals," *High Country News*, July 21, 2008.
66. Graham Holmes, quoted in "Council Studies Power Proposal," *Window Rock Navajo Times* (AZ), May 29, 1969.
67. For Navajo Generating Statistics, see William Brown, "Black Mesa," in *Los Angeles: Biography of a City*, ed. John Caughey and LaRee Caughey (Berkeley: University of California Press, 1976), 457.
68. Environmental Preservation Classes of Many Farms High School, *Environmental and Economic Issue: The Strip Mining on Black Mesa and the Coal Burning Power Plants in the Southwest* (Many Farms, AZ: Many Farms High School, July 1971), 2–3.
69. Rosanda Suetopka Thayer, "Navajo Hopi Group Meets to Discuss Peabody, Navajo Generating Station," *Flagstaff Navajo-Hopi Observer* (AZ), July 12, 2010.
70. Stewart Udall, interview by Jack Loeffler, 1983, in *Headed Upstream*, 122.

Stewart Udall, photo by Jack Loeffler

Jack Loeffler

Politics of the Colorado River with Stewart Udall

THE DEGREE TO WHICH WE ARE SHAPED BY CULTURAL PERSPECTIVES AND mores during the present super-speed continuum of cultural evolution is incalculable. The mind-set of octogenarians seems ponderous to today's youth. Each of the five or so generations spawned in America since the death of John Wesley Powell in 1902 have labored within an exponentially increasingly complex cultural milieu that in today's world moves at the speed of light relative to days of yore.

By the end of the nineteenth century, wave after wave of adventurous seekers of fortune had followed the advice of Horace Greeley by heading west. By 1900, the human population of North America was estimated at 75 million souls.

The tone was set for the new century. Economics dominated the minds of entrepreneurs and elected officials. It was time to use the new technologies that were emerging from the Industrial Revolution of the last 150 years to make the twentieth century the greatest in human history. The United

States of America was now an empire that spanned the continent by right of Divine Providence according to Manifest Destiny.

That was five human generations ago, and the population of the United States is now estimated at time of writing to be 312,652,657. In just 12,000 years since the end of the Pleistocene epoch, the population of our species has increased from an estimated 1 million to nearly 7 billion. And our expenditure of natural resources has grown proportionately.

Bearing all this in mind, it is fascinating to try to imagine the cultural mind-set of a century ago when modern transportation was a train pulled by a coal-fired steam engine, modern media was to be found in daily and weekly newspapers, and modern communications were mostly limited to telegraph wires and the Morse code. Cars, aeroplanes, radios, and telephones were on the immediate horizon, but television wouldn't become a cultural phenomenon till around 1950. The personal computer has now been with us for about twenty-five years, and the Internet even fewer than that. Those of us spawned in the first half of the twentieth century have born witness to enormous change.

I HAVE BEEN GREATLY PRIVILEGED TO SHARE FRIENDSHIP WITH A FEW notable fellow humans born in the twentieth century, that pivotal span of a hundred years marked on one end by horse and buggy transportation and instant worldwide communication on the other. The late Stewart Udall is one of these people. Over a span of more than four decades we engaged in conversation that greatly expanded my purview and allowed me a clear view into the political workings of midcentury America.

I regard Stewart Udall as a man of enormous integrity with an evolved sense of ethics. In my mind, he was our greatest secretary of the interior, who ultimately battled the U.S. government in behalf of Navajo uranium miners and their widows and families after he left public office. He was a champion of Native American rights as well as the natural environment. He was born to a Mormon family in early 1920 and grew up on a farm in St. Johns, Arizona. His father was a highly respected judge who contributed greatly to Stewart's understanding of politics and its role in shaping cultural standards of our country's population. Stewart possessed a refined sense of history and while in office befriended great writers and interpreters of history, including Robert Frost, Alvin Josephy, and Wallace Stegner. Stewart was one of three people who provided me with a profound sense of the role of history. He chided me for reading fiction to change the subject after a long day of conducting interviews or writing or pursuing research.

I recorded many hours of conversations with Stewart over a span of twenty-seven years. The following is excerpted from a conversation I recorded in 2001 when I was producing a six-part documentary radio series entitled *Moving Waters: The Colorado River and the West.*

SU: You need a geographical perspective to really study the Colorado River. It has ended up as the most regulated, the most governed by detailed laws and so on. But the mountain West, the Río Grande . . . in some respects, it's a long narrow river. It goes all the way to the Gulf Coast of course. It's similar [to the Colorado River]. It's a relatively narrow watershed. The Columbia River of course is a unique river. It's the great hydroelectric river. It starts in British Columbia, western Montana, the Snake River. The amount of water in that river dwarfs the Colorado. It is sort of fascinating to me that the Colorado River has ended up historically as being the . . . most controversial, most regulated by law, and in another aspect, because southern California is not part of the river basin, is unique in the sense that the main benefit of the river historically has been California . . .

The interesting thing to me about southern California and the Metropolitan Water District is that it is not part of the Colorado River Basin. The southern California people, beginning with what I think rightly could be called the rape of Owens Valley . . . had audacity. They were big thinkers, and some of them saw that that wonderful climate you have in southern California, not that it had an agricultural potential, although some of the original water went for agriculture, but here you could build a great empire in terms of people. So now you have 16 million people and still growing in that area. When you look at what Los Angeles was in 1880 . . . San Francisco . . . northern California was the hub of California. Los Angeles was a cow town, a cow county, and that was a pretty good description of it. And then the growth began and of course it exploded in the twentieth century. It's fascinating and it's ironical, because it's unique that much of the water to spur the growth and the sprawl and urbanization of California comes from outside the basin from the Colorado River . . .

When the Reclamation Act was passed—big news for the West—in 1903 under Teddy Roosevelt, this set up the Bureau of Reclamation, and there would be projects built in various parts of the West. The Bureau of Reclamation would be running the show to a degree. They would be managing projects and advising Congress what to do and what not to do, although members of Congress ultimately made the opinions. The first big project, really magnificent project, under the Reclamation Act of 1903 was the Salt River Project in Arizona, harnessing the waters out of their mountains. And that project in the Phoenix area is of course still thriving today. It was a wonderful farming irrigation project finished in 1913. But as a result of the Reclamation Act, which was a broad act covering the West, you immediately encountered Indian water rights. And in a case in Montana, because that's where one of the first projects was to be built, the Supreme Court handed down what's called the Winters Doctrine, and that said in effect that when Indian reservations were created, that the law assumed that they would have sufficient water for whatever needs they had then or in the future in their little river basin or river system.

That of course was a godsend to Indians. Not much was done to implement it, but it's been there. That doctrine, I can tell you, in the 1960s when I was there [as secretary of the interior], we saw it as giving up authority to demand that Indian water rights be defined and protected and that Indian projects be pushed forward. We didn't do enough, but that one lawsuit in many ways was more important than the Law of the River [Colorado River Compact of 1922] as they call it . . .

The thing that has distinguished the management and control of the Colorado River from other rivers was the Santa Fe [1922 Colorado River] Compact. I don't think in 1922, when that conference was held, there was any great feeling in the basin because there were small states, relatively lightly populated. Denver of course was beginning to grow. These were mountain states with the high Rockies. That's where most of the water came from. And this was early in the period of building big dams. The dam building was not much on the agenda then although some people dreamed of big dams.

California wanted the [Santa Fe] conference. Herbert Hoover [then secretary of commerce who presided over the meeting] served them well. I've never understood why he was there and what pressure he put on everybody because he was in effect California's agent in putting this through. They wanted an agreement that would enable them to go to Congress and build the Hoover Dam/Boulder Dam. That was a kind of forced marriage. It ended up with this strange thing that never happened in other basins of each state being allocated a certain amount of water. This is purely arbitrary. There was no rational basis for it. It wasn't like the kind of river basin planning that we got started all over the United States in the 1960s when I was secretary of interior, saying now let's look at the river, and what are the highest and best uses; let's try to exercise foresight and look at the future. They just arbitrarily said, and I'm sure Hoover exerted pressure, that they had to have an agreement. That led of course to the construction of Boulder Dam.

That enormous project was then, maybe it still is, the highest dam in the world—went forward in the depths of the Great Depression. This again showed the political power that southern California was exerting and the economic power, because it got the electric power companies involved, and this was in essence a California project. It wasn't anything beneficial to the basin, and the water, other than that flowing down the river to the delta and flowing to the few irrigation districts like Imperial [Valley], huge users of water—the key water, rather good-quality water, was going to southern California through an aqueduct . . .

The Santa Fe Compact simply divided waters and produced an interstate agreement, although Arizona refused to participate, and they fought it for twenty years because they saw it as California getting the upper hand and dominating the river. They were correct in my opinion in that assumption. But they had to have a law, and that became the Boulder Canyon Project, which would spell out what was going to happen if this big dam was built. And that was the beginning really of what they now called—this was a lawyer's dream—the Law of the River, and lawyers in all these states had to become acquainted with the law, and then additional law was written. There's a body of law called the Law of the River. This is not true in any other river basin in the United States, or to this degree. Law governs everything.

JL: In 1944 the United States and Mexico entered into a treaty that talks about how much water Mexico was going to have gotten, because part of the whole concept within the 1922 Compact was that the Lower Basin receive 7.5 million acre-feet and the Upper Basin 7.5 million acre-feet, and Mexico was left out in the cold. So could you talk about what led up to this 1944 treaty?

SU: The interesting thing . . . is Herbert Hoover in Santa Fe and the Santa Fe Compact they created is completely artificial. There's no Upper Basin and Lower Basin; there's a river basin. They divided it up and they left Mexico out. And then the Hoover Dam was built. So I don't know the history of the 1944 treaty with Mexico, but it was during the war, 1944, and obviously [there] was an effort by President Roosevelt to mollify the Mexican officials. Of course they had the agriculture south of the Imperial Valley in Baja California. It was a rich agricultural area, and they wanted to protect their water. They said, 'You divided up the river, but you left us out.' So this treaty, they backed up in effect and guaranteed Mexico a certain amount of water. That was the 1944 U.S.–Mexico Treaty. It wasn't a treaty with the states; it was the United States of America and the government of Mexico.

JL: What's interesting to me is that in 1922 it was thought that between 17 and 18 million acre-feet of water came down the Colorado River in a year, which proved to be a fallacy. It's somewhere between 13.5 and 15 million acre-feet a year. So with the allocation within what was then determined to be the Upper and Lower Basins plus Mexico, the river has become very overallocated and that led to many future events. Actually one of the events that that led up to was the 1956 Colorado River Storage Project Act. I wonder if you could talk about that a little bit.

SU: One of the interesting things and shortcomings of the famous Santa Fe Compact between the states was not only that they allocated each state to have a certain amount of river water and left Mexico out, but also one of the big mistakes they made, because they arbitrarily, based on this rather short history, had to decide how much—they shouldn't have been so rigid, and they shouldn't have made an assumption of this kind. It wasn't rational in my opinion. So they had to back up and consider the fact that the flow of the river was less; that Mexico had been left and adjustments had to be made. The Upper Basin states—this is interesting—it wasn't until almost twenty-five years after the compact that they decided they had to have an interstate compact to govern their relationships and they did this. . . . Congress has to approve it, but they had a compact between the states to set some ground rules. . . . What the Santa Fe Compact did was make what could have been a simple solution a complex solution that had problems, problems, all the way down the line.

JL: The Colorado River Storage Act was what really resulted in the construction of the Glen Canyon Dam.

SU: That came out of the [Upper Basin] states' compact. After World War II, the Upper Basin states, as they were called, decided—California had its big project, Hoover Dam . . . they decided that they wanted to begin projects—dams, irrigation projects, hydroelectric dams—primarily to provide benefits and uses of water. They would tell the Bureau of Reclamation, 'Go study where hydroelectric dams should be built. Go study and tell us how to control the river so we can keep some of the water in our states.' I'm talking about the northern states, the Upper Basin states. And the Bureau of Reclamation would do this, and they put together what was called the Upper Colorado River Storage Project.

JL: Arizona and California have been at loggerheads over the way the water would flow since basically they became states, basically aware of themselves in juxtaposition on either side of the Colorado River. Could you talk about the history of the whole Arizona versus California decree of 1963?

SU: Well, you have to historically look at what became the biggest water fight in the West between big California and little Arizona. Neither of these states contributed a great deal of water to the river. The main contribution from Arizona was the Gila River, which originates in the mountains in Arizona and New Mexico. And yet Arizona, although they walked away from Santa Fe, they didn't agree, they were given a pretty good allocation of water. But afterward when you got to the 1940s, California already had its aqueduct on the river, and Hoover Dam had been built, and Arizona wasn't getting much benefit. So Arizona wanted to have its big project [Central Arizona Project (CAP)], and they had the Bureau of Reclamation study various alternatives, what should be done.

When they brought that to the floor of Congress in 1951 and '52, Arizona senator Carl Hayden, who was a very powerful figure on the [Senate] Appropriations Committee, Arizona then had for two years the Senate majority leader. They passed it right through the Senate, they just whipped it through. California blocked it. That was their strategy. The longer they could keep Arizona off the river, the longer they could use water they weren't entitled to, you see, and that was what created a lot of enmity and a lot of suspicion in Arizona. They [California] blocked the legislation in the House. Arizona had the chairman of that committee. California was throwing its weight around and said no, Arizona's water internally hadn't been qualified, and there were disputes with California. They had to go to the Supreme Court.

Well, that was a nice strategy because it bought eleven years of delay. That was finally resolved by the Supreme Court confirming Arizona's contention in 1963. Then Arizona's project [CAP] could start to move forward.

JL: One of the determining factors within the way the water has been allocated and continues to be used is the concept of the best beneficial use. Could you talk about how that concept has evolved, especially within your tenure as secretary of interior?

SU: The idea of water rights, Powell saw this early, that the West was an arid—semiarid—region. It had high mountains and a lot of water would flow out, but if you're going to use it you had to build reservoirs. Powell was a dam builder. . . . You had to develop projects to use it [water]. Hydroelectric power was a big thing back at the turn of the century and later on. Hydroelectric power is clean. It's a marvelous resource. So harnessing the rivers was the language. But in terms of water use, when you got to the West, unlike the eastern part of the United States, which is water rich and you had riparian water law, the idea was established early in various states, and then it became rather general, that the first person to start using water had a prior right, and that became and still is a central part of this region, which is dry and semiarid and water is very precious. So how do you determine who has the right to use water? Within states, each state had a certain amount of water allocated. They had to work within that limit, but they could decide by state law what the priorities were. They're still wrestling with this in some of the states.

JL: After the Upper and Lower Basins were established within the framework of the Law of the River, the secretary of the interior was granted the *mayordomoship* of the Lower Basin. Could you talk about that?

SU: In working out the Santa Fe Compact, I think it became clear in this river basin [Lower Basin] where water was limited and there would be controversies between states that you had to have an authority who could make important decisions, who could force the states to take certain actions. And they selected the secretary of interior, and this is written into the laws, the Boulder Canyon Project and so on. The secretary in effect became the water master of the river and could make very important decisions. For example, as dry periods developed, you have water shortages, and you're shifting water down the river. As Marc Reisner pointed out [in *Cadillac Desert*], the whole river is a big plumbing system, and releasing water at certain points and certain times became very crucial. The secretary of the interior was given a major role as the water master.

JL: In the Upper Basin it's a different situation. It's my understanding that the states themselves have autonomy in the way the water is used within the states, and the agreements are between the states rather than under the direct control of somebody outside the states.

SU: Naturally when you allocate water to Colorado, to Utah, to New Mexico, it was left to the states to have primary authority to decide how they were going to use *their* water. This is what apportionment meant: that's their water and they can decide how to use it. Still there were conflicts. This is a river basin, the river's flowing. And the states encountered problems between themselves and they had to work this out. In some instances they had the authority, and the secretary of interior simply was there to regulate if there were disputes that arose.

JL: One of the things that became more apparent as time wore on throughout the whole twentieth century, there was the evolving sense that the Lower Basin states, especially California, were using water that was supposed to have been allocated to the Upper Basin states but had thus far gone unused. So within the Upper Basin was a growing concern that California might indeed get more and more of the Upper Basin water.

SU: You see, what happened with the Boulder Canyon Project and the building of Hoover Dam is an established fact. California not only tapped the river for southern California's water use through the aqueduct, but it also confirmed the water rights for the farmers on down the river. But here's a big river system, and this Upper Basin/Lower Basin idea didn't have much rational meaning to the river because rivers don't obey Santa Fe [1922 Colorado River] compacts. They flow. As long as the Upper Basin states didn't have major reservoirs to hold water. . . . There was plenty of water for this incredible delta at the end there [Sea of Cortez]. And Arizona was sitting there watching the water flow by, and the only projects they could get started were a few on the Gila River and some small projects for Indians right adjacent to the river itself.

So naturally the Upper Basin states, they were very loud about this in the 1950s. Well, we're not using our water, and unless we can have projects it all flows down to California. [This thinking resulted in the Colorado River Storage Project of 1956 that shortly led to the construction of the Glen Canyon Dam and the enormous reservoir named after John Wesley Powell. This allows the

Upper Basin states to control the amount of water to be released to the Lower Basin states and Mexico (JL).]

JL: Could you talk about the Salt River Project in southern Arizona in the early twentieth century?

SU: The Salt River Project in my view was not only one of the pioneering western irrigation projects, large scale, but one of the most successful. That dam was completed in 1913. The Salt River Project is an irrigation project. They divided the land at that huge farming area near Mesa, Arizona, and the surrounding area. The farms fed off the canals that came out of the Roosevelt Dam and even on into Phoenix itself and Glendale. Those huge canals were taking this water, and these were very productive farms. What the Phoenix metropolitan area has ended up doing is what California did. For most of the farms, the land was [monetarily] more valuable for developing subdivisions [contributing to] this enormous growth of Phoenix trying to be another Los Angeles. The farmers, it was so attractive they sold out, and they tore down the citrus groves and the cotton fields. You see now this huge glob of urban sprawl. That was one of the most successful irrigation projects. They also developed hydroelectric power and provided power for people. It transformed the Phoenix area, the Salt River Project.

JL: Then the Central Arizona Project—do you think that the Salt River Project provided the model of thinking for the early forms of the Central Arizona Project?

SU: The early concepts—I'm talking about the 1940s and early '50s when they first developed the Central Arizona Project, which Arizona congressmen and senators wanted—the model was the Salt River Project of 1913. And we told Congress—I testified many times as secretary of interior—that this was going to be based on the success of the Salt River Project, and the water being brought in would be used for farming. That was the primary use that we had in mind. Well, by the time that the project was finished, it was so costly, the farmers couldn't afford the water.

And the primary use of the water now is for urban use, and it may end up rescuing Tucson, which has grown too far [large] and too fast.

JL: Could you talk about the actual evolution of the Central Arizona Project [CAP], the history of it and how it gets its water?

SU: Well, Arizona with its water, the portion under the Santa Fe Compact [1922 Colorado River Compact]—see, Arizona walked away. They didn't ratify the compact. We had a governor in 1944 who said, 'Well, if we're going to get a project we can't do it ourselves.' There was a lot of 'We'll do our own project. We'll finance it ourselves.' But that never materialized. So they [Arizona] finally joined the [Colorado River] compact. You had to have a project, and the Arizona people wanted to bring the water into central Arizona, into the Phoenix area for farms, into Pinal County, the big cotton farms, very productive cotton farms. My brother then said later on let's take the aqueduct all the way to Tucson, and it's going to prove ultimately to be the salvation of Tucson, I suspect, as a source of water. By the time this original project was proposed or put together in 1950 and 1951 and presented to Congress, the Senate approved it. By the time you got this huge project ultimately built you're on into the 1980s. And it was very costly, and by then this massive urbanization had begun and the farms were being converted into subdivisions in the Phoenix metropolitan area. And the other farmers who had overpumped the underground water in Pinal County south of Phoenix, the [CAP] water was so expensive most of them couldn't afford it. So the project [CAP] has ended up as an urban water supply project with some agriculture, not a great deal.

In subsequent conversations, Stewart Udall revealed that he was gravely disappointed that CAP water proved too costly for farmers and has ultimately contributed to the too-rapid growth of Phoenix and Tucson. He had grown up in a Mormon family in the small farming/ranching community of St. Johns, Arizona. As a teenager during the Great Depression, he worked part time as a cowboy. During World War II he served as a belly gunner on a bomber and survived at least fifty combat missions over Europe. He ultimately attained a law degree from the University of Arizona and married Ermalee Webb from Mesa, Arizona. Stewart and Lee Udall raised six children.

As a politician, he was elected to the U.S. Congress, where he served from 1955 till 1961 when he was appointed as secretary of the interior by President John F. Kennedy, a post he held for eight years. He worked vigorously in Arizona's behalf to claim the state's "rightful" apportionment to Colorado River water that had seemingly been thwarted by California politicians for four decades, a water war that greatly shaped his political perspective.

Stewart Udall was one of America's greatest environmentalists. During his tenure as secretary to the interior, he oversaw the addition of four national parks, six national monuments, eight national seashores and lakeshores, nine national recreation areas, twenty national historic sites, and fifty-six national wildlife refuges. He contributed to the Clean Air Act, the Endangered Species Act, and other acts designed to protect our environment. It was on his watch that the Wilderness Act of 1964 was signed into law.

Stewart Udall died in the first minutes of spring 2010 at his home in Santa Fe. President Barack Obama stated, "For the better part of three decades, Stewart Udall served this nation honorably. Whether in the skies above Italy in World War II, in Congress, or as secretary of the interior, Stewart Udall left an indelible mark on this nation and inspired countless Americans to continue his fight for clean air, clean water, and to maintain our many national treasures."

Roy Kady, photo by Jack Loeffler

Roy Kady

ROY KADY, NAVAJO (DINÉ) FROM TEEC NOS POS CHAPTER, IS AN ACCOMplished weaver and fiber artist who is breaking out of the traditional mold with new styles. He lives in his family home at Many Goat Springs above the community of Teec Nos Pos, where he maintains a large flock of rare Navajo-Churro sheep and angora goats, plus one llama. He teaches culture classes at the local K–8 school, works with at-risk youth, and recently taught weaving classes at the Diné College, Shiprock campus. With TahNibaa Natani, he organizes monthly spin-offs for weavers from all cultures, which they host in communities within the Shiprock–Teec Nos Pos area. He is committed to participating as a member of the Lore of the Land because for many years he has sought to record stories that are told in the landscape, taking elders and practitioners to the many sacred sites in the vicinity that are linked with Diné creation stories and ceremonial stories. The Teec Nos Pos area, on the eastern slope of the Chuska Mountains, is rich in lore of all kinds, including that of plants, land management, livestock, and agriculture.

Applying Navajo Tradition to the Modern World

(The content from this chapter is excerpted from two interviews conducted by Jack and Celestia Loeffler on April 2, 2010, and July 22, 2010.)

RK: *Tó* [water], in our language, is as important as the sounds that it makes. Like *tó*, it makes the sound of water. The *tch* part is a slow-moving, peaceful stream and a creek that just makes that *tchhh* sound. And then the more rushed forceful part makes the *kwooooh*, the last part of the word. So in that sense, the word is sacred for water, and powerful. It's a life provider, giver, and that tó is something that we just can't do without. As we all know, what percent of the Earth is made of tó? But we're quickly disrespecting it in so many ways and contaminating it.

In my community there are several natural springs where the tó comes out in a natural form to come greet us, to remind us that yeah, "Take care of me, and I can take care of you . . . fertilize and water your corn, your fruit trees, your wild medicinal and edible plants." And so they [the waters] also travel and they also are gifts of both Mother Earth and Father Sky. They're embedded in both of those two entities. And they both feed one another. The evaporation goes back into the father, and in turn when it gets full, it comes back [to Mother Earth] and releases it again. It's a cycle. So with that, we know that it's sacred, it's totally sacred, and that it needs the respect that it needs.

Oftentimes you hear people say, "Oh yeah, we haven't had adequate rainfall, we're in a drought, we need to all band together and we need to pray for more." Yeah, we do, but we also need to respect it more. We can't just pray for it and disrespect it. Somehow that doesn't make sense. You can't disrespect and then pray for more. It doesn't work that way. So I think some of the other things that we're quickly forgetting, some of the offerings that we would give to these waters' natural abilities, need to happen some more. Because again when they feel like they're being neglected and disrespected, they have every means to cease for a short moment, or for a long moment. It's really up to them.

It's interesting too that in the Navajo Way a lot of the things, the natural elements out there, are considered living beings just like us. In some cases they're referred to as people. There's the Earth Surface People, there's the Rock People, there's the Water People, there's the Sky People, there's the Plant People, there's the Mountain People. They're all referred to because they have life just like we do. They're not any different. What they resort to is just like what we as humans resort to—it's really not any different—to survive. So that is something that I want to say again about tó. Water's so important.

It's really important because there's a lot of references to water, not just in our ceremonial structures, in our stories, but even namesakes, like where I'm from. I'm from Goat Spring. Its interpretation in Navajo is "where the goats come to drink." It's referring to the natural spring that used to—it's still there but it's been modernized and it's been encapped with a well system. But it's really an interesting watering hole, because you have to think back and say wow, this is where all the animals came before it was established, before the school was put there. I think why I'll always say, when I introduce myself, that I'm from Goat Spring is because I reconnect to it in that way. I think that is important.

And that's why I want to do that "river of life" for our community, because I know a lot of that sort of connection, even more so the reconnection, will be made because more and more we're just being driven from it in terms of everything becoming too modern. I think that we should still respect those [traditional] stories and teachings and those resources. Some of them [the water sources] dry out for just so many different reasons, I think. There's offerings that are not made there anymore. The usage of it is not respectful. They're for commercial purposes and wasteful intentions. And they [the waters] know that. They have their own set of values and feelings. So that's how naturally they go back to sleep.

They rest, they rejuvenate. But it can be a continuance if it knows its purpose. So that's how we should all be thinking like a watershed and respecting it and using it—just enough.

When I asked my elders, my kin, how was it determined that we were going to move up the mountain? How was it determined that we were going to move in that direction? It was all about water, where the watersheds are. The knowledge and the idea, [from] having been there, they knew there was a water source not too far from the spot where we were going to be. So it's really important, I think, the water, to be thinking like a watershed in that sense.

Water to us, it's a part of the four elements that were put in place for us. It's one of the main elements in life that you can't live without. Just the same way as we say sheep is life, we also say water is life, air is life. Without each one's component, there's no life that exists here.

JL: Is there anything that you can reveal about the Diné point of view about the San Juan River?

RK: The San Juan River, I've heard stories referring to it as a sacred river—it's a sort of boundary in a sense, but the boundary there is also an opportunity as well as a life giver.

But the question that I posed [to an elder] was, "What's the meaning of *tota'a* [for] Farmington [New Mexico]? I know *toh*, but what's the *a'a*? Is it *tah*, meaning three waters that come together? Or is it the gap? I posed the question to an elder, and I said, "What does it mean, *tota'a*?" Then he said, "It's the gap between the three." There are certain things that weren't taken in between these gaps. I think he was referring to the development. And so especially certain ceremonies are not taken beyond these boundaries. They stay within the opposite side of the rivers. Though we can continue our life and pass through, ceremonies specifically do not happen across there. They're not taken there.

Then he also referred to the river as when they dammed it and imprisoned the water's natural ability to be free and to flow free. So he also made reference to how we've captured it without its permission. There's a lot of things that we've lost as a people, and he says that we struggle to this day to prosper in the way that we truly want to prosper, because we've captured that and we don't have the natural flow of the river anymore.

So in that sense the San Juan River also has stories about [itself] because it created the boundary, it was our source of protection and safety. We crossed it when we wanted to barter with the Utes. But when a disagreement or warfare evolved, we were quickly and safely to go back on the other side, because we believe that the water has many powerful attributes for our survival and our safety. So I've referred to the river as being that, and of course the biggest thing is that it's a life giver.

The San Juan River is something of a sacred being that has a life, but offerings aren't being made to it as we have done in the past. I think when it's missing something like that, there comes a point where it just will want to go to sleep until the next opportunity.

I didn't feel like I got a complete answer when I asked the elder. That's always the case when I talk to an elder, because you're given that opportunity [to discover the answer for yourself], and I like that.

I know I didn't get the answer, but now I'm learning to accept the answer that I get. And from there it gives me the ability to search for another section of that. It's another way of them telling you that you have a whole long life here to figure [the answers] out. What's your rush? You'll get it, the answer. It'll come. You may be sitting on your doorstep at ninety-something years old, and [it's] like a light bulb coming on. You're sitting there going, "Oh-kay! That's it!" I think not knowing the answer gives you that motivation, that commitment, that dedication, to do the work. Like for myself, there's a lot of those questions that I put out there, and I work with the answers that I got now. Like for sheep, for instance. Why is it the way it is? If we truly valued [the sheep] and we say it's sacred, why is it more and more that we're not really grasping it and making it what we say it is to us? I think partially the answer for myself is, well, you're the one to continue the stories and the lifeway. And then by doing so you'll get your answer. It may not be now, but it'll be somewhere along that road.

Because I think some people, they get the answer right away, and their idea is okay, good, I'm done. It's now moving on to the next phase. But for myself it helps me to hold on to the sheep lifeway because there's really some unanswered questions for me. And being told that the answer will come, just stick with what I've taught you or what I've answered so far because the rest of the answer will come later. So I think that helps me in keeping with it.

Because there's so many times that people get frustrated. Some of my own peers have shared this with me. I think the other society, the other culture, provides that kind of teaching: "Well, if I don't get the answer, what's the use? What's in it for me? I may just turn in the sheep and maybe buy a couple of cows. Because really I'm not seeing the whole purpose of raising the sheep." I think that that's what happened with a lot of them too, where they do trade it in for not just monetary things but trade it in because they haven't really been given the partial answer that I get. It makes you impatient to live in the other world.

JL: Can you talk a little bit about the Healthy Native Community [Project]?

RK: It's really an opportunity. They give you a lot of tools that are basically geared toward utilizing a lot of our elders' teachings. How did our ancestors, how did our grandparents—how would they have written their vision statements? It's really healthy, meaning everything, not just individual. It's your whole environment. The whole cycle of life is included. And it's basically just getting back to basics in terms of creating a healthy whole community.

We've revived our farming opportunities. That again is the gift of water. The water, it's important in that sense. I hear a lot about cultural and traditional teachings. And I also hear a lot about, well, this is the twenty-first century and it's all about this here. But really, it's a balance of all of that. A young individual posed a question to me: "Why do you want to go back? Why do you want to go back and live without electricity, to live without running water, to live without a vehicle?" And I said, "Well, it's not really to go back because I know we can't go back. It's more to get the intention back. It's really to take those tools that were very effective, that had made our people the way they were, to go back and get them, to bring them up at this time and to utilize them again. Because they worked then, and they can work here too."

Really our people were very conservative about so many things. And do you know why? I make my mom as an example. She utilizes things that we just quickly dispose of because she's come from that life. She knows that life. In fact, when paper towels first came out, she would wash them until they deteriorated. She still does that with disposable plates. She'll utilize them. She'll have these containers, and she'll use them over and over until it can't be used again for storage. So I told this individual, I said, "They did that for us so that we can do what we're doing now."

JL: It's a system of attitudes.

RK: Mm-hmm.

JL: Now we're in this time when we're dominated by sort of the technological culture, and we're dominated by economics rather than other things that we could be dominated by. One of the things that's come through to me loud and clear is that I have never known an indigenous person who's still traditional who doesn't have a deep regard for the sacred quality of homeland.

RK: True.

JL: Western culture has had this huge proclivity for coming in and trying to commodify everything, turning habitat into money.

RK: Yup.

JL: Those are the extremes to me. The idea is how to bring them back together again so that—you know, there is good that comes from so-called Western culture, but it's run away, and there's a huge amount of good that comes with the indigenous sense of how to relate to homeland. So how do we bring those two things together? That's what you're trying to do here.

RK: For sure. The other experience that we had here just recently was during the spring break we were having our meeting in here discussing some projects. There was a little clitter-clatter out here, and then somebody stuck their head in and asked one of the officials to go out. It just so happened that one way or another our kids here has a source where they get their alcohol and drugs from who comes around like an ice-cream mobile that goes through the neighborhood, parks out here, and all the kids come over.

I really commend my community because they came together after that incident, though it was a part of our community for who knows how long. And they really sat down and just by supporting them and giving them that opportunity [to talk] without the formalities that we're pushing upon them. When we get orientated to be chapter officials, it's interesting. All these orientations, they almost like pound us . . . your "Roberts Rules of Order"—it never gives them anymore opportunity to express. And our people are expressive, especially our elders, their stories initially go back. They take us back, and then it becomes current as they talk [their way] into the subject matter. People are no longer comfortable with that, because the mentality now with my peers and younger is this Western concept of "stay on the subject, be prompt, get it over with, act on it, move on." That doesn't work.

So I gave them that opportunity, and we had over two hundred people here that came together on their own, providing them a place and then giving them the traditional setting. So they really got a lot done. They really began to digest what they've really detached themselves from. And now it's on to how do we make this better?

[The problem is] we work from 8:00 to 5:00, and then we drive two extra hours one way, and then by the time we're home what time goes into spending or even talking to our children, teaching them. And we're off again. Weekends are dedicated to doing laundry, getting groceries, taking care of other errands. Leave the kids home, take them, dump them off at the mall. Somewhere we can just take care of these businesses and then come home. On the way home everybody's asleep in the truck because we're all tired. Get home, everybody goes to bed. What little time we have Sunday, mother cooks breakfast. Everybody grabs their own plate, takes it to their room.

We lack teaching our children anything in this time frame. Our elders teach us that we don't fix things by pointing to the direction, meaning the individual who comes out here to do this [sell drugs and alcohol] to our children, it's not going to fix it that we just get rid of him. We don't fix anything that way. But we could fix it by starting at home, by spending more time with our children.

Even when we're together as a family, my mom sits there, and it's like she's with a full home of English speakers. She just sits there like, "Wow, this is strange. My own children and my own grandchildren." They're [the elders]—they feel like strangers. We need to get back to basics and teaching our children about our experiences.

CL: I was thinking about the conversation we had the other day at lunch about traditional food. I was wondering if you could talk a little bit about all of that. Maybe a good place to start, can you talk about the role of corn in your traditional food ways?

RK: Corn is our embodiment. It's like a capsule that represents our body. It's really our capsule. It's your reconnection to what your whole life was about. So corn to us is our soul food, and we also refer to the cornstalk as our tree of life. It represents our life in every way, from when it's planted, it germinates and it roots and becomes stable and it shoots up, and every branch that comes up represents the certain stages where we've learned to talk, we've learned every part all the way up to the maturity. And our reproductive sense, the pollen represents that. And then the corns are our offspring. The color represents, like for white it's male, and then yellow is the female. There's the mixed corn, there's the red corn, there's blue corn, there's gray-blue corn. There are whole stories that relate to how corn was used in bringing First Man and First Woman into creation.

And when I heard that our body is made up of protein, and corn is protein, I'm like, wow. That's why I think storytelling has to come back, because that's where the storytelling is flavorful. It's really important that corn represents us, is something that truly is valuable.

Any time we have a meal, they say that we should have representation of corn in any sense to be part of the meal. It's also an offering. That's why when I begin to find out about GMO [genetically modified] type of corn, it's really scary. It's in the terms of cloning and messing with the true structure of its empowerment, its ability to continue. It's a scary situation. Our people are not well-informed about this thing that's happening.

CL: Do you think that tradition is beginning to disappear? Is it dwindling or regaining strength in your community?

RK: Here's the scary—well, for me it's scary—I don't know if everybody is afraid like me. But most recently we had an in-law wanting to write his dissertation. [He] came to the chapter and shared with us what his dissertation will be about. He wanted to get permission from the community to interview six elders. He shared with us some statistics. He said that now currently about 7 percent of Diné fluently speak Navajo. And I know that's true, because I'm even talking to you in English. Our session in there [the chapter house], I would say about 85 percent, close to 90 percent is all in English—our chapter meetings, except for when I'm doing my part in Navajo. And so with that, I know once you lose your language you lose your culture and your traditions and everything that ties into that. That to me is somewhat scary. And that will include corn is what I'm saying.

That is something that is dear to me, and I always include it as a reminder to my people that we have to talk more and teach more. So I think it's because of that I think we're really being enticed by what's "over there." We see just the physical content from looking at that house and that car over there, but we really don't know the true, deep abscess that it has. That part we're not seeing because it's concealed with that nice shiny red color and chrome wheels. But truly, if you go deeper and beyond that, you see that ugly abscess of tumor. But the way our people thought in the past, like an onion. They got to the core and they thought from within. We need to think that way again. We're more up here [pointing to his head]. We're more brain thinkers now than from the heart. And that's not to blame, but just to say that that's coming from the enticement from the outside culture. And everything plays a role in it, media and any kind of publication, radio. It's really grabbing more and more hold on our youth.

CL: Marketing?

RK: Marketing. Everything. So corn kind of is becoming the next item that is becoming endangered. I know I go to a lot of local, even family parties where you see the picnic. And you don't see the representation anymore. But at times when it's a personal one, meaning my sisters or when we're having Thanksgiving, my mom always asks her if she'll make that [traditional] corn mush. See, that's why you eat that mush when you marry or you have a ceremony is because it's you, and then the two [people getting married] put together. That's why we do that. What you tasted there, if you could taste life, was the purity of it. And that's the representation of sharing that meal. If there was nothing else left to eat but corn, that's how it would be eaten, and you can survive on it. Somebody said you can live on bread, cheese, and beer—that's what it is to us. You can survive with just that. You have your seeds, you can plant it.

Really, corn—and I wish I could show you a cookbook, but we can prepare corn in so many ways. Some are snacks, some are winter food, some are summer food. Some are just in between, all the way, every way around. It's food just for the heart, for the soul. It reconnects you; it keeps you healthy, just complete healthiness. So corn is an important seed to bring into the twenty-first century and to grow it again. That's why I'm really anxious to meet up with my farmers—I've been away for so long—and just to spend the time. I'm happy, I'm elated, I'm just overjoyed by the fact that they're working hard. And it really was just giving them that opportunity and that little support.

CL: Can you talk about that project?

RK: Sure. Our community here, we utilized the runoff from the mountain. We've been given that gift, and the gift comes down into our valley. Then we share it together, and we plant, and we use it as our water source. In that sense we have ceremonial cornfields that we plant. That all refers back. This is where it's really important to continue the language, because all these teachings that I'm sharing with you refer back to the corn way of life and our Beauty and Blessingway ceremonies. It's embedded in our songs and a lot of our ceremonial chants. It refers to us. That's the long life that we envision, that we yearn to live. When we say *Sa'ah Naghai Bik'e Hozhoon*, you're really referring to that as the way of life coming from the corn and all the teachings.

There's a spiral formation that we dedicate a section of our cornfield. And then there's the waffle [pattern]. Those are purposely gardened in that fashion. And that is something that they [Teec Nos Pos farmers] brought back here to this community. Then they're currently going to be doing that this summer and then engaging the youth in these teachings.

For the longest time, and I'm saying for the longest time in terms of not just a little over ten years of our drought, but maybe even going back another six years, where really people just became so dependent upon running to Wal-Mart and getting the majority of their food source. That's one of the reasons why I know our truths have died out. And I think we also got a high increase of diabetes and obesity. I think now they've [the community has] seen those factors, and they know to make it better is to farm again.

JL: Could you talk about the dome?

RK: The dome is a representation of hope. And then the structure and the meaning behind the dome is referenced to our *hooghan* [traditional Navajo house]. Every representation of that dome is related to the farming ways for us. The dome was a gift to the community that was given to us by Jamie Oliver [the celebrity chef] when he was here when we did that one-hour [television] show.

This is where we want to bring both the elders and the youth together in a farming project. The youth, they want to utilize it in the winter as well to grow fresh vegetables for the senior center, for themselves and their families. So my sister kindly donated a quarter acre of that place where the dome is. And that's going to become a demonstration project that will include these particular sacred farm patterns. And then we're going to do no-till drip irrigation, different kinds of gardens. And then possibly we also want to have an orchard in the back somewhere. We need to relook at our permacultural traditions and [use] them.

Here's the great part. You have your whole community and your whole resources, and then somewhere the light bulb goes *ding* and somebody brings in that idea. So we've written to the LDS Church because they're great food preservers. We're going to incorporate some of our preservation techniques and then theirs, using that and learning that here. Then start freezing and/or canning some of our precious gifts that we get from the land so that we don't have to depend on outside places. And then also to create a marketplace for the community and the passersby here at the junction. We've already talked to the local owner at that particular store, who's given thumbs up to utilize whatever there is to create a farmers market.

Then it went even further than that. The youth again talked about some of the things that could be possible by brainstorming, coming to a consensus. They said, "You know, it would be so awesome that we have a café there where we just do as much as possible local. We go over to Minnie Benally and get one of her healthy cows, take it in to get it processed. That's going to be our steak and our hamburgers and whatever else we need it for. We can go over to my corral and buy some sheep, mutton, whatever. And that will revive [the local food systems]. We listened to the elders, because some of them were saying, 'How many years have we been talking about bringing a local eatery in our community where we can sit and have a bowl of stew and where we can have conversation, where we can dialog again.'"

So they [the youth] listened to him. And that's when they came up with the idea that it's going to be the local stuff that the farmers and they also have to help grow. The biggest thing, and I love this part, is that they will operate it and run it. So they want to learn about business. I said, "Yeah, we'll support you. We have a scholarship fund account we need to get creative with. If you're determined, we'll put you through school to manage a business."

We look forward to being [LGA] certified, our local governments, so that these kinds of opportunities are more possible because we don't have the bureaucratic red tape that we have to go through. We make those decisions here. We have our own economic planning and zoning ordinances. We have our own business licensing office. That's something that will really help us.

JL: What does that mean, LGA?

RK: It's just Local Governance Act, where you can govern as a township. You have your town council, whatever you want to call it, and you choose your own form of government. The people choose the form of government that they want.

JL: This is far out. It's been almost two years since you took over the chapter presidency?

RK: A little over a year; what is it, fourteen months? January was year one.

JL: You've done a huge amount. How does the political structure work here within the chapter at this point?

RK: At this point, the officials, our duty and our role mainly is just to have oversight over the two administrative staff. The two administrative staff are Navajo Nation employees, meaning Window Rock is their main oversight. LGSC [Local Governance Support Centers] was passed a little over ten years ago where it was determined by the Navajo Nation council that it might be better that each chapter become their own local government in their township so that it alleviates a lot of the bureaucratic red tape, different departments that we have to go through just to get anything resolved in that matter.

JL: This is decentralized governance.

RK: Mm-hmm. And our government is funny because at the time that that was passed, it was just kind of like, "Woo, throw it out to them." And when it got here it was like, "What do we do?" It's kind of still the same at certain points. But what we've come to learn is, well, we'll do what we can. We'll get things as to how we understand it. And so far we've been good about those decisions instead of waiting, waiting, for someone to come out here to explain it to us further. So we've just kind of reviewed how this system will be implemented. We were delighted that our five management systems were already created here for this chapter, and it's also because we're a resolution government, that people resolve to implement, to start practicing it.

So when we came here we just took that and then we started practicing and we started reviewing and then sharing it with the people, because that's important. It's important to educate them [in] new terms of how to manage your five management principles. I told them that it can be a very good thing, because now it provides a lot of transparency and accountability because that's a part of the procurement and all that in our management system. And also to me it empowers each community to utilize its resources to its fullest.

JL: You know, I heard a phrase about ten years ago. I was working a project out in California, and some anthropologist came up with it. It's "community of practice." And that's what you're developing.

RK: Mm-hmm, yes. That's true, because that's what it is to us, and that's how we've handled it from the beginning. If you were to see my whole standing committee, and maybe someday you'll see, it's totally diversified. And it kind of came naturally that way. It's people I have already known and interact with and seen, that I've asked to come be part of my cabinet. It's funny when I say that. And then when we present—because we've also been asked to present at other chapters because they're looking at us as a model. And the first thing that they notice is the diversity. I said, "And that's really what has made our community also come back together." Because the idea of a chapter house is, "Oh, it's just those old folks over there griping and it's just for them."

But no, the first thing that I did when I came in—I thought that I had enough youth telling me that it's something that I should bring. And that's the first thing that we created here was the youth council. So they have a seat up here with us when we have our meetings. They're part of the process. And the next thing were the farmers. Those were important. And then the others that were already in place were just kind of there, and they came later into the picture. So that way, that's what the community wants. It's really what they want.

JL: It's coming together as a community. You said that other communities are looking to Teec [Nos Pos, Arizona] as a model. Are they really beginning to get into it?

RK: Yeah. Our administrative staff here, I sit with them and I try to get with them at least twice a month if I can. It's just really sitting together and sharing ideas and how things can work better.

JL: Collaboration.

RK: Mm-hmm. I just put forth on the table different items on the plate and they say, "This is what our community wants." And it's true. It's something that I think each community earns.

JL: It's an evolving process, isn't it? It's alive in some sense.

RK: Mm-hmm. And even more so it's like all these orientations that I went to at the beginning, it was just the same repetitive, "We need to, *da da da*." It's like, "We need to?" What do our people think, first of all? Are they familiar with the process? Have you taught them enough and educated them enough for them to understand the possibilities? I'm not trying to say it's not going to work, but I'm just trying to say have you done that already, or is it just another one of those, throw it out to you, you guys take care of it, you deal with it. And what I've learned that what's always been the case is you deal with it, you figure it out, you try. And it really does put a toll on the community in ways where when they want to see something changed or done they go through all this rhetoric about how it can be done, who can do it, that sort of thing.

Great example here. Our chapter has really brought DBI [Diné be Iina, Inc., a nonprofit organization that supports Navajo traditional lifeway] into our chapter as a collaborative partner. The people again, because we're a resolution government, agreed to do that. And we have a whole project here which is going to be a local wool-washing facility that's all energy efficient. It's basically saying, "You can do it. You have the hands to do it. It's always been there. It'll provide you with the green jobs that we all talk about." To us the green jobs are at home. It's not industrialized or commercialized. It's at home.

And it's interesting. When we get visitors up here, they're antsy because we don't really follow the Robert's Rules. I mean, we can do the basic stuff. We don't do the, give them only two or five minutes for feedback. Gosh, as a Navajo elder you don't have two minutes. That's just the greeting right there. Because that's how you connect. And then people come to that consensus together. But you're disallowing that by saying that it needs to be in a quick process. "Your regular chapter meeting should not last any more than an hour. Two hours is too long."

I say, "Well, it's not really up to you. It's up to the people." Even in terms of—it really is about what the people decide together to make something work. They've always been told, "You're going to be here on Monday on this day at this time, and that's because that's the only time I have available because the rest of the time I'm working." That's always been the case. But it was interesting. When we first met, I said, "You guys are going to decide when you want to meet, when it works for you, what time is best." That whole thing. So it was just bringing those basics back to them.

And I think that's one of the reasons also that a lot of chapters don't keep their administrative staff, because they get frustrated too. Some of them are true community oriented and they want to listen to the community. And then there's LGSC or Window Rock from some department that says, "No, you're supposed to do it this way because it says it right here."

JL: It's unnatural to do it that way. This is the natural way, homegrown.

RK: Yes.

CL: What is the youth project for?

RK: Actually there's two different ones. I've heard a lot of the youth talking about how they're forgotten in the process. And a lot of them sharing with me that there's nothing to do here in their own community, that they wish that they could even just have a place where they could meet and maybe listen to music or play some games and get on the Internet after school and any time when they feel like it. That they can call it their home space away from home. So with that I appointed a youth, sort of a director, just to help the youth find themselves in their community, to extract their hidden voices because they do have hidden voices, just like a lot of adults. To sit with them and to dialogue and say, "Hey, this is your community. How would you like to see your community?"

It's been a struggle, just like anything that is new for our community. So it's been a struggle getting them to commit into the youth organization. I had some wonderful help from a person by the name of Christopher Francis who's done a lot. He's kind of laid the initial foundation in getting it started. This will be actually our third attempt, but we're still very hopeful. Last night we had another meeting, so we're constantly bringing in people to share with us different ways of engaging them.

CL: What would success look like then?

RK: Success is going to look like the way they want it to be. I've shared with them over and over, this is what I see, but it is you that's going to make it the way you want it to be. A lot of times what happens is somebody comes in and jumps in and says hey, this is how we're going to do it, this is the way it's going to be, this is the color of the building, this is going to be the design of the building, and here's what I'll put in it. And never is it, how do you want it? What do you want to see in there? What color shall the building be? Where can we put this? How can you be part of this, in building this, in pouring the cement, in getting a paintbrush and painting the building. That to me is important because it becomes your own. You have ownership to it. When it's just put up for you and you have no process in it, some of the things that you would like to see in there are not being put in there, then quickly you disown it because you don't see what you want to see in it.

To involve people, even just to sit with them and to talk to them about it, and then listen to them and draw from them some ideas, really helps in the end final product. They feel like yes, I've done something. They may not have picked up the shovel or the paintbrush, but by sharing with us how it should take place and by including that, they still have ownership in it. So that's what I want to do with them. I want to continue to work with them.

CL: So they're taking part in creating it. That's really important and very special. Do you do a blessing every day?

RK: Mm-hmm. Every morning I do a blessing. Back there are my two bags that I take out. One has white corn for the morning, and then in the evening is yellow corn. So that's my offering every day. And corn, because we're made from corn, because it's our main staple. So offering it is kind of like saying, "What's your favorite food?" It's almost like fasting. You offer it back, and that amount that you offer back is what you won't have, so you sacrifice that and you offer it because you acknowledge and you say, "Thank you for giving me this." So I offer it back again.

CL: That's wonderful that you do that every day to reconnect.

RK: If I don't have my corn meal, I always have my corn pollen in my pocket for travel. Navajos are really—they improvise. Anything they knew had corn in it, they would probably offer it if they didn't have other means. In fact, I've even heard stories that at one time we've had famine, and of course when you're in that state you're reluctant to give. But they had to. They got to a point where they couldn't, so they offered dirt. Because they knew that what they offered, in those spots where they offered, now it's become part of the dirt again. So they just used dirt for the time being. So those connections, they're there. They don't completely disappear. They just become a different matter. It decomposes into a different substance.

But if you make the connection in your prayers and it's coming from your heart, it suffices to your offering.

CL: The intention is there.

RK: Exactly. Totally.

CL: You were talking about the San Juan River as being sacred and that when it was dammed it had been captured without permission. Can you talk about that?

RK: Anything in nature, you capture it and you enslave it, you give it a different intention, you disharmonize that process. The whole cycle of life gets out of whack. So that's the same thing that happened to a lot of our waters that have been captured for that. And don't get me wrong. We capture water for farming. But at the end of our farming, those ponds are usually dry, back to where we started from. We don't capture it for other means, capture it for commercialization, for greed, and so many other things. But to capture anything that is out of nature and to deprive it from its natural ability to be a part of its community is something that to us is not only disrespectful but also can counter in many ways. And I hear a lot of my elders talk how without permission that took place and how now some of our offering places that were very sacred to us are now underwater.

CL: Actually right now I'm doing some research on the Animas–La Plata Project, and I know that that's one thing that's happening is that a lot of sacred sites, even burial sites, they were excavated, and as many of the artifacts as they could come up with were taken away and are now being, I guess, put in the museum. And the burial sites are just reburied in other sacred places.

RK: I don't think that took place with this dam. It was just not considered.

CL: Really. So it's just underwater.

RK: It's just underwater and it's like drowning yourself. That's how I hear them [the elders] talk about what took place. We're all about balance. And to hold those grudges and to carry on those harsh feelings, we kind of try not to involve that in our lives. Now we talk about them, but back then our people were very in harmony and in balance. I think that was the reason why it transpired the way it did, because they weren't as voiceful as we are in this day and age. But still, it's something that one can think about it in terms of how you've drowned the sacred places where we did our offerings.

CL: Do you feel like your people are in balance?

RK: We're still in balance, but the balance to me is a little bit more difficult to achieve because of all the other influences.

CL: Such as?

RK: Such as media, in terms of television, video games, cell phones. Don't get me wrong, I'm a part of that process. But what I tried to do with my life is to accept them and also think to myself, how can I be less dependent [on these technologies]? They're there, and you can utilize them for emergencies, and sometimes it's necessary. But there are times when you could just overuse it.

CL: I think that goes right into something you mentioned earlier, relearning and reliving old lifeways. You talked about the struggle with sheep in particular, coming into a traditional and a modern way of life.

RK: Again, the balance of the two that I apply to my life is something that I can share and say that it works for me. But I can see it working for others. It's just the application as to how they might initially struggle, because I did. It wasn't just all gung ho for me when I decided to do that. I was more enticed with what's "over there." And if you look at my history of how in high school I didn't want to come back—in fact, right after I graduated I didn't come back. I said to myself, "I want a better life. I want a better life out here, I want a job in a high-rise building, I want to wear a suit, I want to wear spiffy clothes, have a car, have a house"—those kind of aspirations. That quickly dissolved, and I'm glad it did. Because when I experienced it out there, I thought, this is what I wanted? Is this really what I want? Is this the kind of lifestyle that I really want to live?

My answer, not right away—it took some other places for me to wander—to give me that answer that no, my life is here. So I came back to it. And that's what I did. Then the application of the two, some of the traditional teachings and the traditional way of life, applying that to your life in a balance, then including the modern way of life, the outside world way of life—just including the bits that fit together in a more balanced manner is what I'm referring to.

Sometimes I get miscommunicated—I get this from some of my peers and some of my people who listen to some of the things that I say. Sometimes they say things like, "You can't bring the old ways back." I'm not doing that. All I'm doing is I'm bringing with me in my life some of the tools that work for me. Those were the teachings that were given to me by my grandparents and my mother. And I accept them in my life today because they work for me.

The balance of that way of life, it attracts me and I accept it. It's truly created a harmonious atmosphere for me, and that's why I stand by them. And that's why I share with the world that this is the balance and sheep is the balance of my life. If I didn't have sheep, I wouldn't be sitting here talking to you. Nor would I have had the opportunity to see all the beautiful people in my community, the youth and the elders. Every single one of them, I would not have that opportunity to be with them. I wouldn't be here. So sheep created that balance in my life.

I still have that opportunity and venture into that society and to be a part of it comfortably. But I can tell you, I can no longer be there for more than a certain allotted time. I always want to come home because I miss my new lifestyle. I say new because it is new. It is new meaning when I came back to it, it may have been about eighteen years maybe or even less, that's when I decided to come home, balance my life, and to really think about why I was put here. Now I know, this is where it's going to be and this [is] where the people, meaning my kinfolks here, are who I was placed here to help—to help themselves, really.

So that's what I'm talking about. It's not something from the old ways that I'm sometimes miscommunicated as bringing that back. No, that's really impossible. But it's just the important stuff of how our people endured and lived off the land. Herding sheep—I can show you my whole portfolio of where it's been successful. And cookbooks. They've [the youth] seen the show with Jamie Oliver. And last summer it was my sheep who took me to Italy to be a part of a world gathering of food communities. Sheep is something that is going to give me that opportunity to probably travel the world. And have fun and have great food and meet awesome people that later you realize that are not any different than you. We all arrived here together and we are connected, our threads are woven together to make this strong fabric.

In that sense it has given me that opportunity to share those experiences back in my community and to say, hey, this is where I've been, this is how the people live there. It's not any different. And really we're struggling together, and really if you think about it it's not a struggle because it lets us kind of go back to the story of how our grandparents used to tell us where there was no electricity, when there was no running water, when there was no trading posts around. If we think about that, what do we call that? Was that a struggle to them? To me it wasn't because they loved the life that they lived. You create your own day, but you have to embrace it and say, "This is the day that I want to create." You always think positive. No matter what people say to you out there, no matter what kind of negative energy might come your way, you just remain positive, you stay strong. And I think my strong abilities come from my mom because she's always been my mentor. She's taught me everything I know.

And then my grandparents, even though I didn't have the opportunity to learn more from them. My grandfather died when he was ninety-six, ninety-seven years old. I pray and I make my offerings every morning to have his dream. Then my great-great-grandmother lived close to there. My [mom] was saying she's about eighty-six, eighty-seven. So thinking about that, I have readjusted my life and balanced it. Because we only have this one opportunity to be in this physical form. I really want to enjoy it, to live my life to its fullest but to always give—to me I see all those teachings that were given to me as the gift, and that gift became ten-times fold for me to give back. So every teaching that my grandfather gave me, I always think about it [in] terms like cinch weaving. Somebody comes up to me and says, "I want to learn how to weave cinches." Then I create a workshop. To me I haven't reached my full commitment in giving that gift back to people who yearn for it, meaning the gift that my grandfather had taught me. It was a gift that I know his intention was, you'll learn it and you pass on that gift to a hundred more people. I'm still working on that every day. Then every teaching that I've learned, that's how I approach it now.

That's something that I brought into my community in the chapter to be the example. Because they did it, and they could do it again if they could just open up to it again.

CL: I wanted to ask you to talk about, to interpret, the commons. Just what is the commons in your purview? For example, this town is your commons or this watershed is our commons. Does that kind of—

RK: Oh, totally. The biggest commons that comes to mind is again through sheep herding where at one time we were seminomadic. My mom has told me stories about the commons of going to Gadii'ahi, which is thirty miles back to Shiprock, and then coming back around slowly in different areas where at times you live with a different commons of people. You put your flock together. Whatever it is you shared together happened there. And there was a central structure that everybody who wished to be there at that moment had access to. It belonged to nobody, yet it belonged to everybody.

But now we totally act different in respect to commons. How we've become that way . . . probably not to put blame on just one particular movement, but a lot of the influences that were imposed upon us, whether it was our government or whether it was the missionary movement; whether it was at the time when the trading post came about in our region.

Some of the other commons are the farms. These farms, whoever got there, started the initial preparation of the farm. Then they'd continue up to their next destination. The next group of individuals that showed up continued the next step, whether that was planting the corn, and so forth. And then coming back around during harvest, you had access to the farm, and you only took what you needed. And then when that harvest completely matured, the whole community came together to harvest it and share equally. Those were some of the commons. Those are some of the commons that I'm trying to reignite. Because I know it's there, they [the community] talk about it. They talk about, "That's how we used to be." And so this farm here is actually a community farm. The youth work on it every day in the morning when they come here. They usually go out in the morning and they hoe for an hour or two before we start our other projects.

And in turn I told them that they are more than welcome to come back and take their equal share of what they put into it. So when harvest comes, I'll let them know and remind them to bring their family here to take their share because it's only right that they do.

And even the ceremonies, those are commons. Ceremonies are for everyone. I've seen it as being adaptable, meaning I've seen [people] other than Navajo people utilizing it now. I think with the respect and feeling from your heart, that this is what will help you, the door is open for you to engage and to practice that. So I think when the elders talk about it, that's how they refer to it still. When you ask a medicine man or when they're talking about this particular topic, they say that the ceremonies were given to *us*. They always make a reference and they say, "It wasn't given to me. It was given to *us* for the sole purpose of recreating the balance and the harmony that we've wandered off of; to help us, to strengthen us, to heal us, to make us a better person."

And it's for all of us, not just me. Actually this is a commons that we share to this day is our ceremonies. Maybe that's what's holding us, I think, because I know the songs and some of the chants and the prayers, they are done in the language. So when I say I think that's what's holding us this far, that's why the language is so important to learn, so that the tradition gets carried. And if we're compassionate and if we're true believers it works, but if we're just doing it for the sole fact of entertainment then we know it's not. It's not a good practice.

So I think things evolve and they become what they become through time. But the main aspects, the main songs, I know they remain a part of the central rituals of these ceremonies.

That's what we're doing with the community farm here and all the other things that we're trying to re-educate our people that it's okay—it was okay then, it's still okay now. We can still do it.

And you know how our current president [Barack Obama] had the "Yes we can," [slogan] I often talk about that too. We did that. Now the next step is "Yes *I* can." We have to do it individually, too, in order for it to happen. Our contribution counts, and there's so many different ways you can contribute.

I was saying it earlier to the kids, how it was explained to me by them. There's different ways to contribute. Even if you can't do the hard labor, maybe you can bring a loaf of bread to feed the workers, or even a glass of water to give some hardworking soul a drink. That is a contribution, isn't it? Every bit counts. Every bit makes it happen, and even just of your word in thanking that person, that's a huge contribution into what we're trying to do. And so they're learning.

CL: Do you have any advice for the future generation?

RK: I think to be uniquely themselves but to also remember that it's okay to apply our history to modern life. A lot of times when you bring that up, it's a giggle. I know they don't really mean it disrespectfully, but to them it's funny because they didn't experience it; they think it's a funny matter when you try to share with them that, oh yeah, this is how it was, or this is how Grandma lived. They just kind of like giggle and think, oh, that's just really old-fashioned, or gosh, we can't still be doing *that*. That kind of attitude. But to tell them that it's okay, that how Grandma lived and what she used to do back then, you could still use that [knowledge] now, maybe adjusted a little to fit the current time. It's very much okay. That's what I've done.

And then just that they are a part of this community, each one of them.

And balance is a really strong word for me, because it's that that keeps me on the right path. I feel like if I lean too much a certain way, I start feeling the aftereffects of it. So I quickly reposition.

And they're beautiful souls, they're beautiful individuals, they all mean so much. They could just do so much. I can see that. But

I feel like they feel that their hands are tied somehow. I often want to just untie it and say, "You're free. Let's go catch your dreams and let's put the foundation down. Let's do it. Let's show ourselves we can do it. That you, one by one, can do it."

That's what I really want them to know is they can, they absolutely can. That it's necessary for them to know that there's no limit. But they need to apply themselves and dedicate themselves and be really true. Just be true and honest about it. And in no time their dreams will be fulfilled. They'll see it sitting right there in front of them, and they'll reflect back, wow. And then they'll teach the next generation that same thing.

CL: Shall we end on that beautiful note?

RK: Wonderful, yes. That's where we'll leave it.

Camillus Lopez, photo by Jack Loeffler

Camillus Lopez

Camillus Lopez is a traditional Tohono O'odham whose culture is indigenous to the Sonoran Desert. He is a lore master and has devoted much of his life to collecting the stories and songs of his people. He has served as the chairman of the Gu Achi District of the Tohono O'odham Nation and has been a Lore of the Land scholar since its inception. Camillus and his wife Mary have become active in radio production for KOHN, "the Voice of the Tohono O'odham Nation." It was the result of a conversation with Camillus Lopez in 2000 that the concept for the Lore of the Land had its genesis. His chapter is excerpted from three interviews conducted by Jack Loeffler over a period of six years. His words reflect the deep and abiding sense of spiritual kinship his people have for the plants and animals, the mountains and arroyos, and the desert sands of their homeland.

Tohono O'odham Culture

Embracing Traditional Wisdom

(The content from this chapter is excerpted from three interviews conducted by Jack Loeffler between 1994 and 2000.)

There's certain places in the mountains that have natural water. The water goes through. It's like these little tunnels go through the mountains, and I guess the water pressure goes up a little bit, and they would take water from that place. There's other places where there's holes where they set somebody to watch them so the cows and the deer and all the other animals stay away from it so that people get that water. I guess they used some kind of a stick, stick it in there and then they would turn it, and as they turned it would go deeper into the mountain, and then water would start coming out and that person could get their water.

People on the flat land, the daughters and sons were asked to run every morning, but it wasn't for exercise. It was to get water. The boys would take their bow and arrow and protect the girls as they ran. But long before in the [Tohono O'odham] culture, the young men didn't talk to the young women, and the young women didn't talk to the young men. So they kind of stayed apart but still protected them, running in the desert. Then they get to where the water is, and the boys do whatever it takes to get the water out. If there's no surface water they have to go down. Then the girls get the water, and then they run back to the house, returning about midmorning, and then they have that water.

But back then one of the big things was to share a lot. So when people ran out of water and they didn't have enough, they would just find a family who had a lot of girls and had a lot of water that they could get some. Everybody's related so sometimes someone would just give water automatically because they know they need it, if someone didn't have any daughters or if the mothers couldn't run. And sometimes if a woman's menstruating they won't let them do stuff like that. So that's how they got their water. They come back, and they use it, and then in the evening time usually it's gone, so then they go back. But when they moved up to the mountain villages for the cactus harvest, there was a lot of water there because then it was closer. They didn't have to go that far.

The belief in O'odham is that the mountains make the clouds. If you're sitting under a mountain or even if you're far away and the clouds are coming over, it looks like the mountains are making the clouds and they're stirring them into the air. So here are the clouds coming over the mountain. In the O'odham belief system every mountain makes clouds. So this is talking about the Frog Mountain, Babat Duag. So this person who sings about this sees Frog Mountain and the clouds coming over and the swallows flying underneath at the base of the mountain, and they're happy because there's clouds. In the desert when you have clouds, it's not so hot because it blocks the sun for a while. So it's a good sign. All these songs that we sing are all about bringing rain.

They're asking the frog babies if one of them can sing the songs, the rain songs. Not singing [in a] human voice but to make the sound because back when, whenever it rains in the desert, after the rain comes then the rain has gone away and it's real humid outside. Then you sit out there, and you listen, and you hear the frogs over in the wash real loud sometimes. A long time ago, for people to hear that was a great magic, a great sound. It was what everybody wanted to hear because then you know it had rained. Frogs talking, singing their songs. That's the magic that brought the rain. So that's heaven, when the frogs sang their songs. So they're saying to the frog children, to sing their song. In other words, make it rain. And the frogs bring rain. They're called After the Rain People. When there's a rain and you sit outside and there's no music and the lights are all off and the sun has gone down and there's just that crimson, and you listen to the frogs at the pond and they're talking to each other, they're talking this frog talk, and it's beautiful to people because it had brought rain. Long before we had faucets and water tanks and stuff, this was the most beautiful sound that people looked forward to hearing. So there's a song that they sing about the frog babies because that brings new life, new songs. So it's asking the babies, the children of the frogs, they're saying that maybe you might know a new song. Maybe you might know a song to teach us. So why don't you sing your beautiful music?

The green rainbow. They call it the green rainbow, but it's all the colors. Green because the green is the one at the bottom. You start with red on top. Green is at the bottom because a long time ago people liked to see green because that meant that there was moisture there; it had just rained, it was going to rain again. When everything is green it means that it's going to be bearing a lot of food for the people.

When they say *Pima*, a lot of people talk about Akimel O'odham being the river people. They're not really desert people, but I guess they are now that the river's not running. But back when the river used to run they were Akimel O'odham, because in Gila River region it's the same tribe as us, they have the same stories that we hear here, and they have their own dialects. So they're O'odham. O'odham people live around there, to the north, south, and west of that mountain [Babat Duag] and along the river.

In Gila River [with Akimel O'odham] when I heard the stories there it wasn't so much emphasis on water. It was more emphasis on the river, to keep the river flowing. Then in Tohono O'odham you hear more emphasis on where I'itoi [an aspect of the Great Spirit] did this to bring the rain. And then there's Hiacite O'odham. They have the same myth that we do; it's just that they had their own section of what was O'odham—I think west of Ajo area, that way.

I'itoi is a small man. I don't think he's like a dwarf, but he's small in size and with long hair, kind of grayish but long, like an old man, but sometimes he can change. He has greenish skin. They used to chase him to catch him. I guess it's like the leprechaun thing, kind of. In the songs they called him *juh-deg* I'itoi. *Juh-deg* is the word for green.

So that's how I imagine him. There's a lady that told me that I'itoi came to her and tried to take something from her. So she went and kicked him in the mouth, and she said that he's this little green man, and I went and kicked him in the face and then he ran away. I don't know what happened to him. There's other stories when they would say, little green man, he's running around and he's helping people or he's doing this.

His personality is like, he creates the trouble, and then he sits back, and the people come to him and ask him to fix it so he becomes a hero, even though he's the one that did whatever. He's a human being in a way, but he's got [a] special gift to have life if he had to. People had killed him in the myth. There's many stories about him, how he does things. He comes and smokes with people, he comes and drinks with people, and he gets drunk and he sings the songs. In the myth he's very human. I'itoi was the one that created human beings. When we look at life it's not this whole thing about this one God that is good, and then you have the other side which is bad. That's the newest thing that has come in. But in our culture, there is something that has the ability to do good and bad, eventually turning it into something good.

You mimic the movements of Nature. In O'odham there's no categorizing, no science, math language. I noticed that in O'odham culture everything is cold or hot or dark or light or whatever. But in the Western culture there's a lot of numbers and degrees and miles and those things. When in O'odham something's far, then it's far. If it's cold, it's cold. Because I hear a lot of times with Westerners that I work with, they talk like if it's cloudy they'll say something like, "Oh, this is strange weather for June." In O'odham that's the way it wants to be, then go with it. So everything's based on the outside because it's affecting the inside. When you live as part of an environment, I guess you become it, and everybody else around you is it.

In O'odham long before, you weren't walking through the desert to take a walk. You didn't walk through the desert to go from here to there. You actually were meeting all these trees. So it's actually like a high reverence where you could stop and pick a berry or whatever from a tree, and at the same time you're talking to the tree, saying you've been standing here and these are good for me and I'll use them right. So it's not where you're just taking a walk. You're always remembering that things are alive and that you're not alone.

I walk through the desert, if I see a palo verde tree, I look at it and see that the leaves are yellow, they're pink or whatever color the leaves are, then I think that that's the time of the year in that way. So you recognize the tree as you would recognize a person walking by. Two O'odham in the middle of the desert coming together and they start talking; they're not talking to themselves. You're being overheard by all this other stuff around you. So there's always been that kind of reverence. If you see a palo verde tree standing there and you come by, you notice it. Then it's like both people know where you're at. So it's not just singing or talking to the trees as you're walking along but keeping in mind that you're there with them and that they're just as much alive as you are. I think trees sense more of that than humans do.

In O'odham culture women don't necessarily go looking for power, for strength, because they have so much strength already. They have a womb where babies are made. That's a lot of power. When they go into menstruation they used to go out and get away from men because they might make them sick.

Men are always seeking power. That's why men do the vision quest, that's why men do all these things, looking for the answers for whatever, for strength. They're the ones that go to war. So it's those kind of songs that can give me some kind of strength, I guess; empowerment to push. A lot of people talk about Baboquivari, where I'itoi lives, with great reverence. There's a great power there. That's true, but to me it's these mountains that are here in Santa Rosa, these mountains that I remember.

Each place has a place in the natural order. To do something with that place, just to take something out from the natural order would cause disturbance to the rest of the order. I guess like the river, how it kind of went into the ground because the farmers were taking from the water below, and the river disappeared in Santa Cruz. That makes the water table go down. Therefore it doesn't rain so much here because there's no natural water coming from the ground evaporating into the sky to cause the moisture that we need up there for the clouds to come through in the same way.

You build something somewhere, the respect. You have to remember that there's an order there. You go somewhere and you take a rock; you don't replace the rock with anything there, you disturb the balance. That kind of a thing, you know.

The mountain holds a special place in history or time. There's a reason that it's put there. Nobody owns it, it owns itself. In O'odham, it's a strange thing to own land because the land was there for everybody. It was placed there by I'itoi to serve a purpose so people could live there and do what they needed to there. You take over and you call it yours without the respect that it should have. It was there before. When ants are living in a place and you're coming for a picnic for one day, you put out ant repellent. You destroy the ants, you destroy the natural flow. So when the ants are gone, you need to replace it with something, but you're just taking off and the ants are gone, and you've only used the place for one day or for a few hours. You don't think of what's going to happen ten years from now.

This place has been here for many, many thousands of years, and human beings have only been here for ten thousand, twelve thousand years. Then in just, what, fifty years, all of a sudden all this stuff is happening. There's no regard for the land.

You try with respect to get it to the way that you found it with an attitude where it's not there for just you, but it's there for all, not just people but for all living creatures. What you do there is going to affect everything else. What you do to the land you do to yourself, and eventually it'll catch back to you. You have an attitude that you want this place to be here with the respect that it had for all these years with the great power that it has. If you want the power it's there, but you have to be open to it. If you're there for yourself, that's okay, but you have to remember that other people, other living creatures, are in this life with everybody else. If they can respect, it's not hard for someone else.

To learn a thing, you have to learn to respect things. If you don't respect things it's just because you never learned it somewhere. So to those people that are there for themselves, they have to learn how to respect things. Because if they don't, it affects everybody.

Community is everything. It's the stars. It's the ground way under. It's the little ant that comes across. It's Coyote. It's the buzzard. The actions and stuff, it reflects who you are. And if you can see yourself in it, then you're there. But if you can't look at Nature and see yourself in it, then you're too far away. That's what I think one of the things people need to do is go out and become, look at the mirror of Nature and try to see themselves in it, because if they can see themselves in it, then they can help themselves by helping the environment.

There's Baboquivari. That's where I'itoi lives. That's what's sacred about it. But in the O'odham culture every mountain is sacred. Every mountain has its story. When you share a song about that mountain, you don't say it's any better than that other mountain. There's no levels, no degrees to more sacred or less sacred. It's that every mountain, every little sand, every wash is sacred. It has a story behind it. That's why it was made. If there was no reason for it to be there, it would not have been made. So every mountain has its sacredness, its reason for being on the nation. Here on our nation there's a lot of little spots that people don't know about. What we like there is the children's shrine. It's a story about four children that were sacrificed to stop the flood a long time ago, and it's still there. It's got different reverence to it. Then there's Ka'a-dak where the olla sits, and that's what the word *Ka'a dak* means. And it's connected to the children's shrine story. Therefore everything around Santa Rosa has to do with the children's shrine, but it's not just Santa Rosa. I'm talking about regions, not just villages, certain communities.

One thing that I noticed too is that some cultures will highly revere sacred places. They'll build these majestic castles and buildings on them. In O'odham and in most Native cultures, all the sacred shrines are kind of hidden. They become part of whatever the area is around there. So if I asked you to find Ha'a dah-kin, you'd probably have a hard time because it doesn't look anymore grand than what's there. In O'odham we say if you're lucky you'll find it, and then you'll get your blessing.

We have also Ventana Cave up there. There's a time clock up there. There's some red markings on the wall. When the sun comes up, the shadow hits that certain part, and you know what part of the year you're in. I guess that's how people used to know when to plant and when the rains were coming and when there was going to be a drought and stuff. And the moon follows that. If you follow that clock, it's real easy.

There's Naoo-ch'haf-binee. There's a clown that ran during ceremony and he died there, so they just buried him right where he lay down and died. There's a place where the turtle shrine is at. They have these turtles that they have put there for some reason. It was regarding rain. They put the turtles there and there's seashells and stuff in there. It fell a few years ago. A person came and took the stuff that was in there. Then we have a little lady. There's a little place up past Hickiwan. There's a lady sitting there, a little rock. Then another rock behind it, a big rock with holes in it. The young men, when they do their ceremony, would come and pick up those rocks and offer something to the little lady and sit back and throw the rocks. Depending on which hole it fell into, that was predicting your life, what your life was going to be like—whether you were going to be lucky with women, if you were going to have a lot of crops or a lot of children.

They have that lady sitting there with her dog, and she's making a basket, and every once in a while she gets up to go put more wood on the fire. The dog comes in and unravels the basket. So they say that if she finishes that, the world's supposed to end. There's the place up towards Ga'ak where they have this man laying there. They call it San Francisco. It's out of rock. If you stand at a certain angle you can see it. Then you have way up Superstition Mountain the people that were turned into stone back then long before, on top. They were sent up there because the floodwaters got so big they would send somebody down there to check to see if the water had gone down there yet. And the water hadn't gone down so they'd go back and forth, and they'd keep sending somebody. So the dog was sitting there and just said, "Well, we'll check and see if the water went down," and said it in a human voice, and everybody turned to stone, and then they've been up there ever since. So you go up there, you see them.

There's a place where an old woman born around Kitt Peak who ran up that way, and she lived in the mountain and she started killing the children, eating them. So they had to find a way to destroy her. And there's a long story in the myths of how they killed her. And they took her and put her bones up there. And they had this big dance in honor of her before they killed her. Their dance floor is still there. You go to Oso Verde about an eighth of a mile north of that village, there's that dance floor—big dance floor. Then you go behind, towards the south there, and that's where they buried her.

When the first flood came and destroyed the people, I'itoi had created the flood. So he made an olla and put himself in it. And Coyote knew he was going to die too so he made a flat raft with reeds tied on. And Turkey Buzzard flew up to the top of the sky and grabbed on to the top. But all the other birds, they grabbed on to the sky and they hung up there. And Tcuwut Makai [the Earth Maker] went down into the ground, underground. So when this big rain came it flooded the whole place, and then they were floating there for many days. And I'itoi and Coyote had made a pact that whoever came out first would be the one to be called the elder brother. So they floated for many days—forty days, forty nights. That's what they say, but I don't believe that. I think they took that from the Bible and Noah. But after the water had subsided they felt that things weren't moving. They had slept and weren't moving anymore. That woke them up.

And Coyote came up first and I'itoi came out at Baboquivari Peak. Coyote found himself somewhere else. I'itoi made a home there at Baboquivari in that place where he had landed. So that's how his house was. So soon after he decided to go look for Coyote. Coyote was looking for him for a while, and he was yelling out, calling out to see if anybody else, if I'itoi was still alive. And then way over there I'itoi heard it, and he called back and then they kept calling back, then they met someplace. And when they met at that place, I'itoi had pronounced himself to be the elder brother even though Coyote had made it first.

But I'itoi doesn't talk like us. He talks backwards, you know. I'itoi was a medicine man. I think that the great thing about him is he's one of the creators. He was the one that created O'odham and Hohokam, the ones who were here before. And he's the one that was the great medicine man of all. He had the greatest power. But he's not treated as apart from O'odham. In the other cultures they have their gods as way out someplace. In O'odham, I'itoi lives among us. He's one of us, and he's still here. There's little myths and little stories that are told about him doing stuff even today. There's a lady that I met just most recently who said that I'itoi came and tickled her heel, and so she turned around to kick what she thought it was, an animal or something, and here is I'itoi. So she chased him out to catch him and he took off and she couldn't catch him. Little green man ran away.

But he's one of us. He's not removed from us. He makes mistakes like we do. He has human emotion, everything.

The way that he's described is as a man, little green man, greenish in color, and he has long hair, long hair streaked. Because the reason he's green and the way it was explained to me from my grandfather was that when kids play in the grass they'll get up and their knees will be green, their trousers will be green where they're playing in the grass. Because they were playing on the grass it turned their trousers green. But I'itoi, when he did stuff, people would chase him down, and he's a little guy so he ran through these bushes really fast and they slapped him. And as he's running real fast, they started turning his skin into a greenish color because that's how he travels is he runs fast through the bush. So running through the bush turned him into green. In the songs they call him [sounds like *chutuket I'itoi*]. So therefore he's green. Baboquivari became where his home was.

Further down where the petroglyphs are, when the rocks had fallen, they had fallen a certain way. The man in the maze metaphor is hewn in that.

The way that you see the man in the maze now is with the opening on the top. It's upside down according to my grandfathers. It's supposed to be the other way around and then come into it with the opening on the bottom, it's your life. You're going into it. Now that first journey is a straight journey from the opening of the hole to the middle part. And the middle part is where the death spot, the place where death is. But in O'odham when you say *mu*, that means death, but it doesn't mean the end. It doesn't mean you stop. It only means that something else is coming after that. So you follow that entrance into the first part. That's where you're in the womb. You're living in your mother's womb—until you hit that black spot. Now that black spot has always fascinated me because long, long before I had asked one of my grandfathers what death looked like. I was always asking questions of different people, and I asked him what death looked like. And he was telling me to go in the rest room and look and stuff. But I remember there was a mirror there. When I looked in the mirror there I saw myself, and I thought, I was looking around for death. I didn't see death, so I forgot about [it].

And then a few years later I saw the movie *Return of Billy Jack*. And in there he had gone for a vision quest and went into this mountain where he found himself. He was looking for death, and he found death, and death looked exactly like him. So I remembered that, and years later I had an experience with death, and it brought back all those memories, and I finally realized that's what my grandfather was saying, is that death is an exact replica of who you are. So to look at death is to look at yourself just like your shadow. Your shadow does what you do. You can't lose it. And when you die you lay down and there's no more shadow—consumed.

So when I studied that and I looked at that metaphor, there's four times in your whole life that you're going to hit that metaphor. That first time is in that place where you go out and then you hit that part where you meet yourself. And then there's the first turn when you turn towards the other way after the spot, and then you start moving into the maze.

So then you follow the longest path from then to the next time you see yourself again, and that's where you learn how to feed yourself, learn how to get along with people, you learn how to do stuff that will take you through the rest of your life. Then the next time you see yourself in there is the time when puberty happens. It's scary for people to go through puberty.

In O'odham they had ceremonies. So they do the ceremony to kill the little boy so that the true man inside could live, and that's where you get responsibilities, and then you have to become an adult and lose your child ways.

Then you move into the next place in the metaphor where childbearing happens. You see your child that's supposed to be a mirror of who you are—actually two people put together, but you're supposed to see yourself in that child somewhere. And therefore it's your responsibility to raise the child and to make the child become who they were meant to be, not who you want them to be but a combination of the two people making another one. And that's whatever the child is called to be and whether the parents like it or not.

Then you have the last part, the journey of where you go and you end. You come one more time before you die. You hit that point again. And when you hit that point, this is where you turn around and look at your life. Most people, I think, get to that point and then look at their whole life and say, "I've made my mistakes, and I've tried, and I made some good stuff go on, and now it's time to see what's next." And then you go into that little place there, the area that's true death that starts life over in a different form. That's where you truly really die because you're living in a place where there's a big dance, the dance of life, going on. And you have all the food that you need. You don't have anything to worry about, and you're just enjoying life.

But if you never learn how to do the dance, which is the dance of love and giving and sharing, then you don't know how to be in that life. Therefore there is hell, because the definition of hell is basically isolation. You isolate yourself from the community. But you need other people to do stuff with. You can't do it alone. So that is basically the metaphor of the man in the maze. If you draw that man in the maze, you can just kind of see where you're at.

Throughout the myths there's a lot of death, and in the metaphors they say it's a continual thing. So in O'odham they don't talk about a heaven and a hell like in the Christian sense. In O'odham they don't talk about a heaven and a hell. It's a dance, the way my grandfather described it. It's a dance. There's the great dance and you go and you join. If you've danced the dance of life, you'll know how to dance the dance of death.

There's fruit trees growing all the way around it so if you get hungry you go and eat fruit, and then you go back to the dance. And some people sing. So you do the dance, but the idea, I think, of the understanding of hell in the O'odham sense, a lot of the Native sense, hell is like isolation. Nobody isolates you. You isolate yourself. If you've done things in this life that I guess are not very helpful to the community, not helping other people, more selfish-type things, you're isolating yourself by removing yourself. The Great Power never leaves from you. It always stays with you. You're the one who decides to leave. So therefore after this death then, after you've met your death, in a way you have to kind of atone for that. And in a way if you wander on your own, even loners need to see somebody just to be inside of someone. In O'odham, the way I've understood it is that you walk for a long while without being with anybody, and in a way you're atoning for your own self. In a way, death, for people that haven't really been good to the community, good to the whole, would be like wandering for many years until someone can find them to bring them back in. Again, they've isolated themselves so people go out there and they don't find you; it's just because you don't want them to find you.

So in O'odham there's no heaven and hell, but there is a concept of loneliness and isolation versus being with family and having a celebration. To me real heaven and hell is not going to the fire, and there's this guy with the red underwear running around with his fork, and he sticks you every time you do something. And there's no idea of this totally good person. It's the idea that totally good things come out of your relatives and your family, and the bad things come out of your selfishness and your wanting to do things without community, without people. That's why communities are important all the time.

They say that in death things are green. In O'odham the green represents the green trees and lush green places—everything's lush and green where the deer and the horses and the cows can eat and be filled. It comes off a little bit of how people used to go out and get deer meat and stuff. If the deer eat a lot, then they're big and they're fat and people get a lot of meat off, and the birds and everything, if there's a lot of food out there for them to eat. If there's a lot of grass and everything's green, the deer eat the grass and then the bobcats eat the deer. So they take the bobcats, stuff like that. It's like that cycle of life. And green, when it's cool. The springtime, you can walk in cool places and it's not hot. They talk about the north wind coming, and it's nice and cool. Sometimes when I used to walk with my grandfather or used to sit outside sometimes and the rain would come, but it would be like a slow drizzle, and he would say, "This is what heaven is like," or, "This is what the afterlife is like—nice and cool," because long before air conditioners and swamp coolers, long before that was how you wanted it to be all the time: nice and cool and not this real hot, hot, hot. So that's what they say. And then it's wet enough to where sometimes you don't even need a drink of water. You don't feel hungry and you don't feel like bloated. That's what O'odham should feel.

But in a way in this life it's different from the next life in that in this life it's more like luck. If you get lucky you put your crops in, your corn and your watermelon. If you get lucky, then you'll eat again. But if the corn doesn't grow, then in O'odham we've always said we'll take what we have and make a life out of it, which is different from the death. Death is more like you have everything that you need, and now it's time because you've gotten to that point where you don't need so much that you come back and help your people that are still in this world. You bring them luck. You bring them rain. One of the things that I've heard about old people when they pass on is they tell them, "Go call the rain." They're supposed to bring the rain. If they get lucky and they find the rain, they bring it there when it's ready.

A lot of these older people—it'll be cloudy, sometimes it'll be rain after they've finished the burial. It'll rain. My great-grandfather is one of those. He knew he was going to pass away. And that day was kind of cloudy, and then when they were taking him to the cemetery the clouds broke and there was all this sun—real nice during the funeral. It wasn't cold, it wasn't hot. Then this dove showed up there, and it went straight for the casket. It went to the cross and stayed there for a while and then it disappeared. Nobody saw it fly off. We did the burial, we finished the burial and put all the flowers and stuff on there, and we started going home. Then it rained a real nice rain. I remember that real well because I know he was one of those people that was really in touch with himself and the culture. And that was the message, go call the rain. And he did.

In all the O'odham culture, Earth, water, fire, air are the only four elements that were born before I'itoi, before Jawed Makeh, before the Coyote. But all those four elements go into the makeup of everything else, and the corn, there's a story about how that came to be. The tobacco, there's a story about how that came to be. There are powerful medicine people that were buried, and this is what came up.

This story about the *hashen* saguaro cactus. It started off where there was a woman that had an addiction to playing in the game *toca*. She loved to play this game, and wherever she went she would play this game. She was one of those wild women, you know. They run, wander the desert, and totally belong to anybody. And they just kind of live alone and do whatever, and they're just wild. They're not connected to anybody—no husbands—and some women had chosen that.

So this woman is real good. She's the perfect MVP [Most Valuable Player] of toca, and everybody invited her to play with their community because again she didn't belong to anybody. So she would live in one place, and then if they were having a game they'd call her to be on their side. So she'd go. Then she gets pregnant at some feast or something and then she has this child. Some stories say the child is a boy, some stories say the child is a girl. But this child is born, and she has to take care of the child. She's taking care of it all right, but she's always playing this game.

So one day she hears there's a game way far across the other part of the land. And she knows that if she goes and she takes the child, she has to carry the child, it would be hard for her to get there fast. She wanted to get there at the start of the game. And the other thing was that if she let the child run beside her, she would still be late. So she had devised a plan that she took four gourd ollas and put some mush in there and told the child, "Go to sleep. When you wake up you take these four gourd ollas and come follow me." So the child got up and started walking to wherever she thought the mother had gone. So the child walked for a while and then got lost. So she saw an eagle there and went and asked where her mother is. The eagle said, "Yeah, I know where your mother is, but I'm not going to tell you until you give me one of those things that you have there." So the little girl gave one of the gourd ollas to the eagle, and the eagle said, "Come follow me" and then flew away. And the little girl followed the eagle wherever it flew to, but it flew so high that it disappeared, and the little girl didn't know the direction. So she lost her way again.

And she saw a hawk sitting there and came to the hawk. "Do you know where mother is?" The hawk said, "Yes, I know where your mother is, but I'm not going to tell you until you give me one of those things that you have." So the little girl did the same thing. The hawk said, "Come follow me" and flew as hawks fly. The girl started running after it, couldn't keep up with the hawk. It was going too fast. And then she lost her way again and came up to a crow sitting and asked the crow, "Do you know where my mother is?" And the crow said, "Yes, I know where your mother is, but I'm not going to tell you unless you give me one of those things that you have." So the little girl gave the crow one of the gourd ollas. So the crow took it and said, "Come follow me" and then flew off. She followed and couldn't keep up with the crow either. So she got to a place where she saw the mourning dove and went to the mourning dove and said, "Do you know where Mother is?" And the mourning dove said, "Yes, I do, come follow me" and then flew from tree to tree as mourning doves fly, from tree to tree—not high up but where she could follow. If she lost her way he would coo in the way mourning doves coo.

She followed the mourning dove all the way to the playing field. She went and sat at the sidelines for a while, and the mother came and she said, "You stay here, play with these children." She took a feather and tied it in the child's hair. She gave it to her as a promise and took off and went to play the game some more. The little girl missed her mother and was looking around, and nobody really knew what happened to her [the little girl], but she disappeared. And then somebody had seen her go to a certain place, and she went into the ground. She was going into the ground because of her sadness. The people saw her and everybody tried [to] catch her. I guess it's like quicksand going down. These medicine men came, and one of them put their arms in there, and they said they just touched her, but they couldn't reach her. She was going down faster than they could catch her.

So they called a badger and asked the badger to go down and follow the little girl. The badger dove in there and followed, digging after the little girl, and got down there. That badger came back out and said they couldn't catch her, she was going too fast. "But I do have something, a little piece of a reddish little thing with black dots," and [the badger] threw it on the ground. Well, the people didn't know what to do with it. So the birds had come and they started eating it. When they ate of it they got drunk and they started flying all directions, no order in their lives. A few years later this thing popped up, this plant. The saguaro had started coming up. Then the people were saying that's that little girl that disappeared some time ago. So they went and they were watching it. Soon, the kids were coming. They had sticks and whipped the saguaro.

So the saguaro got sad about that and then went someplace else. Soon after, the people were asking different birds to go look for the little girl, and they couldn't find her. The Coyote said, "I'll go look. I go running all over the land." Coyote couldn't find the little girl. Then it was the crow that realized that it had seen her someplace, so it told the people, so the people went over there. And the birds all flew over there, and they found the saguaro, the cactus plant standing there. And the birds had come, and they had flown there faster than people, and they noticed that she had the saguaro cactus fruit. So they said, "We better take this before they take it and do something with it."

And they took the saguaro fruit, the seedlings. All the animals said, "Take it and put it someplace where it won't grow."

So they took them to the turtle. They said, "You have your own home, you walk slow, but you go far" and gave the turtle the seedlings and told the turtle to go out to the ocean and throw them in there. So the turtle took the saguaro fruit seedlings and started walking towards the ocean. The people had come there, and they saw that this is that little girl. So they recognized why it had disappeared, and they told the kids, "This is why you shouldn't be hitting the saguaro even though it looks different." So the turtle is walking towards the ocean, and after a few days the Coyote caught up and asked the turtle, "What do you have in your hand?" And the turtle tried to keep walking and tried not to listen to Coyote because he knows that he's sly. He'll do things to destruct or disturb. So the Coyote kept asking the turtle, "What do you have in your hand?" And the turtle kept saying, "Don't worry about it." The Coyote got the turtle to show him. So he opened his paw up, and he saw the little bit of the saguaro seed, and Coyote said, "I can't see it too well. Could you open your hand up some more?" He kept saying, "No. No." And then after a while Coyote kept begging him. So the turtle's hand kept going wider and wider, and there was saguaro seed in there. The Coyote kept saying, "I can't see it, I still can't see it."

The hand was open full all the way. The Coyote hit the turtle's paw on the bottom, and the seed flew up, and it flew, and the wind came and took it. And the north wind blew the seeds to where the saguaro is planted now, and the birds heard about it and they couldn't do anything about it, but now here's all these plants that were going to grow, and people were going to take the fruit and make wine. So the birds made a ceremony, and that's when they made the ceremony for the wine feast and said, "This is how you drink. When you drink, you recognize your relatives, and you drink to bring rain, and you always sing for it. You don't just drink it." So that's why they had the ceremonies, to bring rain so that the crops will grow for the people.

So the older people revere the cactus; it's not just a plant, it's another person. And you're never alone when you're walking in the desert because there's all these cactus that are growing around, and so they're with you. So you're never sad or you never feel alone if you're in the desert. We're taught that these are people, and we try to teach our kids not to put knives and not to mark them up with whatever, their gang signs or whatever—not to mark the cactus up because that's another person, and to do harm to that cactus is to do harm to yourself or to your people.

That ties in real strong with what I was saying—in the O'odham philosophy is that good things can come out [of] bad things. The Coyote did this, and it was bad that he did it, but at the same time it produced something more for us to subsist on. So even I'itoi does things. He'll send trouble, but he sends trouble so that people can come and ask him to fix it. So he comes to fix it. He first does it and then he becomes the hero. So in a way that's kind of like the O'odham mentality.

If you boil everything down you have only Earth, fire, water, and air left, so everything is connected to everything else. There's a song that we sing at the wine feast that talks about just the sun going down. And people are starting to dance, going around in a circle. That's that feeling of euphoria, it's the feeling of oneness where you're just dancing and there's nothing else that matters. All they're thinking about is this music. They're just singing their song. It's one of the beginning songs.

Then there's the medicine people we call—the birds and the snakes and everything. In our creation stories we have four creators in a way. Three of them created people: Tcuwut Makai, this little bug; the *ba'at*, which is the Coyote; and I'itoi—the Creator being I'itoi, he created the people, [was] not looked upon as a good person, a good man. He was a man that did good things, but he also did bad things. He would be the one to send the flood, and then people would cry to him, and he'd come and fix it and stop the flood, kind of like that superhero. In O'odham kind of the way I'm looking at it, is just that life is good, basically good, but once in a while it causes some little things that aren't—and the same way looking at death. That eventually it's all good. Death is your constant companion. But it's not like something to be feared and like that, which I think a lot of people not knowing this other way they don't see it. It's totally different. There's a way to accept it. There's a way to recognize it.

Nowadays, kids go to school and they don't live that lifestyle. There's a few that still remember the culture and know the language and stuff, but there's a lot more that don't. The only way to really know it or to know how it goes is they're born into it, have somebody that's willing to sit down and talk and really show you how these things are and show you the stories. See, you can tell stories about Coyote and all those beings, but to them it's like fairy tale. To me it's like you teach children about Santa Claus or the Easter Bunny or the Tooth Fairy, and then they find out it's not true. I've always seen it, so I've always lived it. But a lot of my friends, to a certain point they got to where they lost I'itoi. Now they're trying to get it back.

Juan Estevan Arellano, photo by
Jack Loeffler

Juan Estevan Arellano

Juan Estevan Arellano is a journalist, writer, researcher, graduate of New Mexico State University, and a fellow of the Washington Journalism Center. He is now a visiting research scholar at the University of New Mexico's School of Architecture. He has received an individual fellowship from the Ford Foundation, and in 2007 the New Mexico State Legislature recognized him as one of the fifteen top Hispanics in New Mexico. He is also involved in setting up the International Acequia Documentation Center, under the auspices of the Lore of the Land. He is a former director of the Oñate Cultural Center and at present resides in Embudo, New Mexico, with his wife Elena. He has served as *mayordomo* and commissioner of the Acequia Junta y Ciénaga and is a former Concilio member of the New Mexico Acequia Association. He is an advocate of traditional agriculture and acequias and is very involved in preserving the genetic diversity of the fruit trees and food traditions that came up the Camino Real de Tierra Adentro from the Middle East via the Iberian Peninsula and Mexico. He is also a board member of Sustain Taos. In the pages that follow, Estevan Arellano reveals the wisdom of generations of Moorish, Spanish, Basque, and other forefathers of modern Nuevomexicanos who attempted to provide guidelines for cultural sustainability in an arid landscape where privatization of common waters was unthinkable.

La Cuenca y la Querencia

The Watershed and the Sense of Place in the *Merced* and Acequia Landscape

In laying the groundwork for developing an environmental history of *La Raza*, land grants and acequias have to be understood, at least in New Mexico. And of course to connect the dots we have to start at the watershed level, at the micro level, before expanding to the full basin as in the Río Grande–Río Bravo binational expanse. This paper will focus on the land and water use in the Río Arriba Bioregion, which extends from La Bajada south of Santa Fe to Colorado's San Luis Valley.

Whether we call ourselves Hispanos, Chicanos, Nuevomexicanos, *manitos*, *paisanos*, or *raza*, there are several important documents we have to consider to fully understand where our concept of a sense of place, or *querencia*, originates. For not everything comes from custom and tradition; that is, some of those customs and traditions have their roots in certain important documents. One might say our philosophy originated orally but has since gone back and forth between oral tradition and the printed page. Among these important documents are *Las Siete Partidas* (1265), compiled during the reign of King Alfonso X (known as "the Wise"), which were not fully implemented until the *Ordenamiento de Alcala* in the fourteenth century; the *Ordenanzas de descubrimiento, nueva población y pacificación de las Indias*, promulgated by King Philip II (1573); the *Recopilación de leyes de los reynos de las indias* (four vols., 1681); and the *Plan de Pitic* (1789). They all laid the groundwork that helps us today to define querencia.

Probably the most important idea that had to be kept in mind when finding a place to settle came from the *Ordenanzas* of 1573, numbers 34, 35, 39, and 40; and also one might argue that this is what eventually defines our querencia. At least it sets the framework, the outline on which to look for that kernel that is querencia, or what anchors us to a certain landscape.

> In order to populate those areas which are already discovered, pacified and under our mandate as well as areas which might be discovered and pacified in the course of time, the following sequence should be adhered to: choose the province, county, and place which will be settled taking into consideration the health of the area which will be known from the abundance of old men or of young men of good complexion, natural fitness and colour, and without illness; and in abundance of healthy animals of sufficient size, and of healthy fruits and fields where no toxic and noxious things are grown, but that it be of good climate, the sky clear and benign, the air pure and soft, without impediment or alterations and of good temperature, without excessive heat or cold, and having to decide, it is better that it be cold.[1]

One might say that these ordinances were the first green laws ever written. Too bad we stopped following their advice, and are now trying to catch up. These laws applied not only to the settling of urban spaces but also rural habitats. Cities that followed these ordinances were Santa Fe, Albuquerque, Tucson, San Antonio, and Los Angeles, among others; also the Villa de Santa Cruz de la Cañada and la Plazuela del Embudo.

Taos, when it was established, followed the model of the *Plan de Pitic*. Known as *Santisima Trinidad del Pitic*, it became present-day Hermosillo, in the state of Sonora. The *Plan de Pitic* was embodied in a royal cedula, issued by the viceroy of New Spain, the count of Revillagigedo, on November 14, 1789, as the basis for the founding of the "new town of Pitic, in the Province of Sonora." It was thereafter the model for practically all of the Spanish and Mexican villas or pueblos subsequently founded in Mexico, New Mexico, and Texas.

The following ordinances, one can say, are an indication that not all settlers were looking for gold, as the majority were looking for a place to settle and raise their families.

(I used the translations by Zelia Nuttall [1921, 1922], revised translation by A. Mundigo & D. Crouch [1977], for the translations pertaining to the *Ordenanzas* of 1573, which were also revised by the author. Translations for the *Partidas* and *Recopilación* were by the author.)

> And they should be in fertile areas with an abundance of fruits and fields, of good land to plant and harvest, of grasslands to grow

> livestock, of mountains and forests for wood and building materials for homes and edifices, and of good and plentiful water supply for drinking and irrigation.[2]

What they were looking for is exactly what they found in northern New Mexico, as they found old and young healthy people living in the pueblos, with plenty of grasslands, forests, and a plentiful water supply from the Río Grande and its tributaries. To Native Americans, the Río Grande, or the Río Bravo del Norte, is known as *mets'ichi chena*, in Keresan, signifying *Big River.* To the Tewa people it is known as *posoge*, or *Big River.* And to the Tiwa it is known as *paslápaane*, also *Big River.* In Towa it is known as *hañapakwa* or *Great Waters* and to the Navajo or Diné, *tó ba-ade*, meaning the *Female River*, because it flowed south.

> The site and position of the towns should be selected in places where water is nearby and where it can be had to take advantage in the neighboring towns and properties, and that materials which are essential for building should be close; and there should be land also for farming, cultivation, and pasture, so to avoid excessive work and cost, since any of the above would be costly if they were far.[3]

> Do not select sites that are too high up because these are affected by winds, and access and service to these are difficult, nor in lowlands which tend to be unhealthy; choose places of medium elevation that enjoy good air, especially from the north and at midday, and if there are mountains or hills, these should be in the west or in the east, and if there should be need to build in high places, do it in areas not subjected to fogs; take note of the terrain and its accidental features and in case that there should be a need to build on the banks of a river, it should be on the eastern bank so when the sun rises it strikes the town first, then the water.[4]

Ordinances 39 and 40 were incorporated into Book 4, Title VII, Law I, of the *Laws of the Indies* of 1681.

Regarding the translations, they had Ordinance 39 reading, "And where it would be possible to demolish neighboring towns and properties in order to take advantage of the materials." The original Spanish does not say anything about demolishing neighboring towns; that's why I've included the original, for those who read Spanish. The original refers to the water being close by, "adonde tengan el agua çerca y que se pueda deribar para mejor se aprovechar della." That part, taken from Ordinance 39, reads as follows in Law I:

> They shall try to have water close by that it can be conducted to the town and properties, distributing it if possible, in order to make the best use of it.
>
> They shall try to have the materials that are needed for buildings, farmlands, cultivation, and pastures, so as to do away with the considerable labor and expenses that result when the materials are far away.

From Ordinance 111

> We order that the ground and vicinity that is to be settled shall be chosen in every way possible as being the most fertile, and abundant with pastures, firewood, timber, metals, fresh water, native people, viability for transportation, and accesses for entry and departure. There shall be no lagoons or swamps nearby in which poisonous animals might breed, or where there might be pollution of air or water.[5]

Once the land grant was made, the settlers had a specified time to move to their new place and start making improvements as noted in the law that follows.

> That possession be taken of divided lands, within three months, and it be planted, with penalty of losing them. All settlers who received land will be obligated within three months, after being assigned, to take possession and plant all boundaries and adjoining lands with willows and trees, and with time prepare all the soil and have it in good condition, some of which can be used to harvest firewood, that they maintain, with a penalty that if time should pass and nothing is planted, the land will be lost, so that it can be provided or given to some other settler, so that space be available in terms of land, but also in the village and ditches that are within every city or town.[6]

By having in a sense a blueprint as to the type of landscape to look for, this created a certain prototype of individual, one who could survive and over the course of time develop a certain mode of character that today we call *corrioso*. An individual that is corrioso is one who in today's lingo is resilient to whatever that person encounters. This also might have led to a certain *mestizaje* since the new settlers were looking carefully at the people already inhabiting a certain landscape, those with "good complexion, natural fitness and colour and without illness," which eventually they mingled with, giving birth to a new individual. Mexican philosophers called this individual *La Nueva Raza*. At the same time this type of landscape undoubtedly produced certain crops and animals that influenced the cuisine. Later this will show how the cuisine became part of what defines our sense of place, or querencia.

Those who had recently come across the Atlantic were used to eating a certain diet, one that was certainly very different from what they encountered in the Americas. The Mediterranean trilogy of wheat, grapes, and olives was very different from the Mesoamerican trilogy of corn, beans, and squash. When don Pedro de Peralta was given the orders to settle Santa Fe in 1609, each settler was supposed to get enough land for the following:

> Dos solares para Casa y Jardin y dos suertes para guerta y otras dos para viña y olibar a quarte caualleriaas de tierra y para el rriego de ellas al aqua necessaria . . .
>
> Two lots one for a house and kitchen garden, and two long lots for gardens and two others for a vineyard and olive grove and a quarter caballería of land and for their irrigation the water necessary . . . [7]

Little did the people in Mexico City know that grapes would not do that well in Santa Fe and that olives would never grow at 7,000 feet. As a result, pig fat, *manteca*, supplanted olives and thus changed the diet of the people who settled in northern New Mexico. Instead of olives for a snack or for cooking, our cuisine became one of *chicharrones*, and the pork meat changed the way chile, another New World food, was prepared. The cuisine became *mestiza* itself, for though chile was from the Americas, pork was a new type of meat that when added changed the cuisine forever. Chile got a different flavor and it produced new flavors, as in the tamales that now had pork.

This landscape "of good temperature, without excessive heat or cold . . . better that it be cold" produced a certain type of individual and also influenced what could be grown, giving rise to a particular lifestyle and cuisine. All these started to mold what today we know as our querencia, because this type of climate influenced who we became as noted by our music, poetry, and folklore.

In both the *Ordenanzas* and the *Recopilación*, *nuestros antepasados* (our predecessors) were laying the foundation for what Kirpatrick Sale calls *bioregionalism* and we call *querencia*:

> The crucial and perhaps only all-encompassing task is to understand place, the immediate specific place where we live. The kinds of soils and rocks under our feet; the source of the waters we drink; the meaning of the different kinds of winds; the common insects, birds, mammals, plants, and trees; the particular cycles of the seasons; the times to plant and harvest and forage—these are the things that are necessary to know.[8]

And probably no one understands querencia better than those who were born on the land grants and use the acequias to grow the food that provides for their sustenance. That communion with the landscape ties us to the enduring code of brotherhood (cultural and historical DNA) just as the poet and musician make the landscape itself the carrier of memory.

By Way of a Definition

Querencia, by way of definition, is someone anchored to a certain physical space and memory. And like the Chinese boxes, you start with the bioregion, in my case the Río Arriba, that space from La Bajada south of Santa Fe to the San Luis Valley in what is today Colorado, but in terms of history and spirit is *la Nueva México*, or *the other Mexico*. That's how the ancients referred to New Mexico in the 1500s. Therefore, Ciudad Juarez and El Paso are also part of la Nueva México, for the same group of people settled those lands.

First, let's attempt to sharpen the definition of *querencia*, for I have seen several definitions, some of which totally miss the mark, I think. I go back to Covarrubias and his *Tesoro de la Lengua Castellana o Española*, the first dictionary of the Castilian language, published in 1611 in Madrid. He defines *querencia* as, "Término de cazadores, es el lugar adonde el animal acude de ordinario, o al pasto o a la dormida," or "A term used by hunters, the place where the animal spends his time, either where he goes to eat or sleep."

The *Diccionario de la Lengua Española de Real Academia Española* defines *querencia* as, "Inclinación o tendencia del hombre y de ciertos animales a volver al sitio en que se han criado o tienen costumber de acudir," or "The inclination or tendency of man and certain animals to return to the site where they were raised or have a tendency of returning to." For our purpose it also means *affection*, *longing*, or *favorite place*. But it also implies a sense of responsibility to that place, a particular ethic toward the land. It is place that people say, "Conoce como sus manos [He knows like his hands]." This definition then defines a mestizo landscape; that is, a new landscape. This landscape can be neither European nor in a sense Mesoamerican but rather a new one, a Nueva México landscape, a landscape that is the other Mexico.

Querencia is that which gives us a sense of place, that which anchors us to the land, that which makes us a unique people, for it implies a deeply rooted knowledge of place, and for that reason we respect our place, for it is our home, and we don't want to violate our home in any way. We like it pristine, healthy, and productive. Our philosophy is one borrowed from our Native American brothers, for we are brothers and sisters: "We do not inherit the land from our parents; we have it borrowed from our children and grandchildren." When my first grandchild was born, it meant I had an added burden, to take better care of the land. Now I have to plant new raspberries and other fruits, like I did when my daughter was born, and that same year she ate fresh raspberries. That is now my task!

Querencia is a place where one feels safe, a place from which one's strength of character is drawn, where one feels at home. And what makes us feel more connected than what we eat, what today we know as "comfort foods."

Even the bull in the bullring prefers a certain place within the plaza where he fixates his gaze and to where he will retreat once he is wounded to rest and feel safe.

In today's globalized world, where immigration has become the norm, especially from the impoverished countries of the south to the northern industrialized countries, as illustrated in the fabulous photo book by Sebastiao Salgado from Brazil, people are losing their sense of place. From 1993 to 1999, he turned his attention to the global phenomenon of mass displacement of people, resulting in the internationally acclaimed books *Migrations* and *The Children* published in 2000.

In the introduction to *Migrations*, he wrote, "More than ever, I feel that the human race is one. There are differences of colour, language, culture and opportunities, but people's feelings and reactions are alike. People flee wars to escape death, they migrate to improve their fortunes, they build new lives in foreign lands, they adapt to extreme hardship . . . "[9] That is exactly what happened when the first settlers with European blood running through their veins made it all the way to Ohkay Owingeh (San Juan Pueblo) following the Camino Real de Tierra Adentro from Zacatecas in 1598, their last leg in a journey that for some of our ancestors started in the Iberian Peninsula in 1492.

Yet there seems to be more disconnect now than ever before in this country, especially as what's happening with the anti-Latino, anti-Mexican, anti-Brown backlash disguised as protecting "our borders." However, to those of us who are of Spanish-Mexican descent, there is no border, for who came here first as "illegal aliens," if not the Anglos who settled in Texas in the 1830s and then ten years later in New Mexico? But we have adapted, and thereby forged a new identity, as Westerners, for we all are foreigners in this land, with the exception of the Native Americans. And if we go back far enough in time, even they might be foreigners if they came across the Bering Strait from Asia. But regardless of how we identify ourselves, we are anchored to this land and place we call home.

In the old days, people in my village and other villages greeted each other with "Buenos días le de Dios," or "May God grant you a good day." Then usually, especially if they didn't know the person, the next question might be "¿Y su merced?" or "And your land grant?" People wanted to know where they were from. I remember my father always asking, "Who's your father?" If he didn't know the father, then he would ask, "Who's your grandfather?" It was not only a way of making conversation but also of grounding himself, of wanting to identify with the person he just met. Usually he would say, "I met your father, or grandfather, back in the '30s or '40s," whenever. But immediately he knew something about that person; he was anchored in time and space.

Part of what anchors us to the landscape that is New Mexico, or the bioregion of the Río Arriba or Upper River, is the way we work the land, the land described earlier by a series of laws whose origins go way back to the Romans. They were influenced by the Arabs as we will see and eventually grounded on the landscape found here.

Today, in every corner the talk is about what is "sustainable agriculture," "organic farming," "permaculture," "subsistence agriculture," "natural agriculture," plus quite a few other terms that are not everyday concepts.

Yet, all these concepts have one thing in common—they are all, in one way or another, derived or based on "traditional" agricultural methods. *Permaculture*, which is a concept known by more people than any of the other terms, means nothing more than "permanent agriculture." It is traditional agriculture, with an $800 spin for a one-day workshop in Santa Fe. It has become a big industry, and though derived from the poor, uneducated farmers throughout the world, it has now become a trend that only those who are educated, sophisticated, and have a ton of money can afford, though they might not have a clue of what working the land entails. Working the land is hard labor, or as the elders say, "es trabajo muy duro." That's why not too many become farmers; working the land is not as romantic as some make it. If traditional agriculture is not permanent agriculture, then I don't know what is. And when talking about "permanent" agriculture, it has to be defined as traditional agriculture, for that is its origin.

Nothing defines this center of "origin" as does the watershed, the *cuenca*, that bowl-like landscape, like both hands put together that resemble an open shell. Our watershed is the Río Grande, and it waters Colorado, New Mexico, and Texas in the United States and stretches from southern Chihuahua to also irrigate Nuevo León, Tamaulipas, and Coahuila in Mexico. But for this paper our focus will be on the Río Arriba part of the watershed already described.

To understand traditional agriculture in northern New Mexico, especially among the Indo-Hispanos, one has to go back several thousand years, and several continents, from the first spring-fed irrigated gardens at Jericho to the *chinampas* of the Aztecs (mistakenly called *floating gardens* by some) in Xochimilco. We cannot talk about traditional agriculture without exploring the European, North African (not to mention Asia and India), and Mesoamerican connections. Though on the surface, when trying to find the Indo-Hispano agricultural traditions, many times they simplistically and erroneously are referred to as *Spanish* in origin, meaning, Castilian, Christian, European. Nothing could be further from the truth because the "old Christians," *los castellanos*, were not accustomed to working the land. Their land holdings were known as *latifundias*, big expanses of land mostly for livestock. Here in New Mexico prior to the Pueblo Revolt these huge land holdings were known as *estancias*. Land grants didn't take hold until after 1695 with the arrival of don Diego de Vargas. Meanwhile the Moors made the Iberian Peninsula bloom and produce new and exotic fruits and vegetables like it had never done before. This is evidenced by the work of many scholars, including Dr. Zohor Idrisi, who wrote *The Muslim Agricultural Revolution and Its Influence on Europe.* But instead of huge pieces of land, the Moors divided the land into smaller, more manageable pieces, known as *minifundias.*

This created a new type of agriculture and resulting cuisine. In terms of traditional Indo-Hispano agriculture, it starts with the land grants, which are then broken down into the commons, acequias and *suertes*, or irrigated spaces. But to be more precise we have to understand the *huertas* (fruit and vegetable fields), *jardines* (gardens, i.e., the garden of paradise—a Persian concept), and *milpas* (cornfields), which is where we find the Roman, Moorish, and Mesoamerican traditions all combined to form a hybrid "traditional agriculture" that is northern New Mexico agriculture to the core. I call this form of agriculture "agricultura mixta tradicional mestiza."

It's *mixta* because it combines fruit trees, vegetables, and legumes; along with livestock, fowl, and bees. This mixed (mixta) type of agriculture developed in New Mexico around the land grants made by Spain and later Mexico, which included both irrigated and dry farming, as well as grazing; growing not only fruits and vegetables but also animals such as *churro* (sheep), *corriente* (cattle), pigs, and turkeys.

And it's *tradicional* because its roots are organic and sustainable because it has sustained Indo-Hispanos for centuries. Since it adheres to the old methods, it follows the precepts of permaculture. Here the acequias and terraces anchor what is the permanent agriculture of this system.

It's *mestiza* because it's a hybrid system of Old World and New World systems (acequias and chinampas), techniques (*surco* and *tapanco*), fruits (cherries and *capulín*), vegetables (lettuce and tomatoes), and animals (chickens and turkeys). New Mexico's agricultural tradition is a blend of Moorish (Middle East and Far East) and Mesoamerica (Tlaxcalteca) agriculture with a sprinkling of Roman knowledge. Therefore, the concept of mestizo combines fruits, vegetables, and methods from both sides of the Atlantic. It's Roman in that the term *huerta* is a Latin concept, *hortus, ti,* from the verb *orior,* "about to be born"; that is where the vegetables and fruits are born and raised. *Milpa* is a Mesoamerican idea, the place where corn is planted (from the Náhuatl *milli,* a seed bed, and *pa,* in this case a "place where corn is sown"). Chinampas were also used to grow corn; they are nothing more than big beds constructed in the middle of the lake made out of mud dug up and piled into beds. Thus they work like hydroponics, the way water is absorbed. If the chinampas are too low they would get flooded, and if they are too high, the plants wouldn't get enough water. The method of watering a huerta, a Roman concept, is by using an acequia, which is Moorish.

An *almacigo* (a place to start plants early, usually inside the house as my mother did) is also part of the agricultural vocabulary in the Río Arriba, and it is Arab in origin. A perfect example of our mestizo heritage is the following sentence, something a father tells his son without thinking about the etymology of words: "Agarra la pala y haz un tapanco en la cequiecita." When we break down the sentence, we find out that *pala*, or "shovel," is a Latin word whose roots are Hebrew. *Tapanco*, from the Náhuatl *tlapantli*, is "a heap or pile of dirt," and *cequiecita* is an Arab word, which means "a small canal to transport water." That simple sentence uses words in Castilian, Latin from the Hebrew, Náhuatl, and Arab-Sabaeno.

The Agricultural Revolution of 1598

With the coming of the settlers under the Spanish Crown, which included Tlaxcalteca Indians and other European nationalities, plus mestizos from *Nueva España* (Mexico didn't exist yet as a country), came a lot of different fruits and vegetables that had never been seen in this part of the world. When don Juan de Oñate settled in Ohkay Owingeh in 1598, which he renamed *San Juan de los Caballeros*, the settlers carried with them their favorite seeds, plus they also brought with them sheep, cattle, pigs, chickens, and horses, revolutionizing the agriculture of the area.

Along with the different plant seeds and trees they also introduced different irrigation techniques and work instruments such as the plow, which forever changed the agriculture and diets of the Native Americans.

Until the arrival of the new European and Mesoamerican settlers, the main staples of the Native population in the Southwest had been corn, beans, and squash, which they planted. Of course they also hunted game and foraged for wild plants and fruits. As is widely documented, within a month of Oñate settling in Ohkay Owingeh, he had 1,500 Indians and new settlers digging what is believed to be the oldest acequia in New Mexico on the Chama River in what is now the community of Chamita.

Oñate knew that in order for the new settlement to thrive and survive, water was needed. According to Dr. Tomás Martínez Saldaña, agricultural historian from Mexico City, who has done a lot of work on the role of the Tlaxcalas in the settlement of the northern frontier of New Spain, it was the Tlaxcaltecas who were instrumental in helping lay out the acequia systems in what is now New Mexico. To be fair, irrigation was also taking place in the Southwest, as evidenced by the great canals uncovered near present-day Phoenix.

But in order to understand the agricultural revolution that happened after Oñate's arrival on July 11, 1598, we have to go back and look at the Muslim Agriculture Revolution that came to the Iberian Peninsula starting in 711 and lasted until 1492 when the Moors were expelled from Spain and Columbus had his first encounter with the so-called New World.

The encounter with America has been the most important marker on the adaptation of new crops, which resulted in an enormous exchange of species between the Old and the New World. Nowadays it is estimated that 40 percent of economically relevant crops originated in America, a fact that sometimes makes it difficult to imagine Old World culture and gastronomy without many American-originated crops. For example, corn, sunflower, potato, tobacco, peanut, cocoa, beans, squash, pumpkin and gourds, tomato, capsicum pepper (simply known to us in New Mexico as chile), and many others that originated in the New World were "new crops" in the Old World a few centuries ago. On the other hand, many Old World crops adapted well in America, and this continent has become the main producing area for them. Examples include coffee from Africa and Arabia, and bananas from Southeastern Asia, as well as soybeans, oranges, limes, sugarcane, and salad greens.

As a result, when the Moors were driven out of Iberia, Castilian agriculture hit a tailspin, as the Old Christians had no idea how to grow food since they had been used to mostly dry-land farming or *secano*, instead of irrigating to produce food. The Tlaxcaltecas were also exceptional farmers in their own right, but they quickly adopted the new techniques, vegetables, and fruit trees to their environment.

Just as the Moors had introduced the new technique of grafting to the Iberian Peninsula, this technique was also introduced to the Americas. Here in New Mexico the people of old started grafting the new varieties of apples, peaches, and apricots by using *trementina*, sap from the piñon trees. They also used (and I still use) mud from black earth when a fruit tree is damaged in order to heal the wound. The Muslims also brought revolutionary social transformations through changed ownership of land. These ideas were later incorporated into the *Ordenanzas* of King Philip II of 1573 and later on to the *Laws of the Indies* of 1681.

Manzanas, or apples, were first introduced to the Americas by the Spanish colonizers and to Spain by the Arabs, who knew them as *tuffah*, in the tenth century. The Spanish also introduced the *albaricoque*, or apricot (known in Mexico as *chabacán*), along with peaches, known as *durazno* and *melocotón*. Watermelons, or *sandias*, and melons, *melones*, were also brought to the Americas from across the ocean. In fact, when the *pobladores* or *vezinos* first came to New Mexico they found the Indians already growing peaches, melons, and watermelons. Until the arrival of the settlers, the Native Americans here had no sweet fruits, so these fruits preceded the settlers as the Indians traded for their seeds in Central Mexico.

Today most of the germ plasma introduced by the early settlers has been lost, but there are still some places along the Camino Real where old seed varieties and heritage fruit trees survive. According to Dr. Martínez Saldaña, the Río Arriba Bioregion is the only place in New Mexico where some of the old germ plasma still survives. Other places along the Camino Real where some of the old fruit, vegetable, and animal species; traditional agricultural techniques; and acequia irrigation traditions still persist include Valle de Allende on the Río Florido in southern Chihuahua and in Bustamante in the State of Monterey, close to Nuevo León. Oñate and his settlers spent two years in the area of Valle de Allende starting in 1596 and planted the same seeds and fruit trees that eventually made their way to the Río Arriba Bioregion.

The people of the Río Arriba Bioregion have been very creative and inventive in times of scarcity, as evidenced by oral history. When there was no coffee to be had, people would roast garbanzo beans and substitute ground garbanzo powder for coffee and drink it in its place. They would do the same when chocolate was hard to come by. They would also roast barley, grind it, and drink it as if it was chocolate. There was always ground blue cornmeal from which the settlers made *atole*, as their descendants still do today.

The Merced and Acequia Landscape: The Anatomy of a Land Grant or Merced

To understand the land patterns in New Mexico one has to understand the history of the *mercedes*, or land grants, under the Spanish Crown, and later the Mexican government, from 1598 until 1847. Though there's been a lot of "chatter" regarding land grants since Reyes Tijerina's infamous 1967 Tierra Amarilla Courthouse raid in northern Río Arriba County, very few people understand the land grants and the struggle of our people.

In trying to comprehend traditional agriculture, what we have in northern New Mexico is a lot more complex than what appears on the surface. Traditional agriculture can never be understood without fully comprehending the division of the land based on the mercedes. There were also *mercedes de agua*, or free distribution of water for irrigation, but these were not as common. We will look at the three main components that constitute a merced. First are lands known as the commons, or *ejidos*; second are the acequias; and the third (with their rigid design that separates the first from the third) are the suertes (because they were allotted to the vezinos, or settlers, by lottery or luck) that are irrigated by the use of acequias. Each acequia forms a separate terrace. If the suertes are the body, the acequias are the veins that give life to the high-desert landscape, and this produces in turn a certain type of cuisine. When you have nourishing food based on grass-fed meat from the commons and fruits and vegetables watered with fresh stream water, then life is abundant and healthy. Dr. Tomás Atencio calls this life, "una vida buen y sana [a good and healthy life]," to which I would add, "y alegre [and joyful]." When people share what they have, it's called *el convite*.

The Commons: Los Ejidos

"From the beginning of the colonization process, communal lands in Spanish American cities became important resources not only for economic ends but as land reserve that made it possible to cope with the population growth that took place in the second half of the eighteenth century," says Carlos A. Page. Page goes on to say, "The Spanish accumulated a firm experience in settling reconquered territories from the Moors, which permitted them to apply their politics to the New World. As such all lands incorporated in America belonged to the Monarch. All property titles came from the Crown through the governor, who had the authority to assign public and private spaces. Among the first were the commons or *ejidos*, which included the lands for the town council, *propios*."[10]

To find out what is considered *common*, we start our journey in the middle of the twelfth century, when King Alfonso X, known as "el sabio," or "the Wise," issued the *Siete Partidas*. The *Siete Partidas* have been cited in several examples of case law, from Louisiana to California, when it comes to water law and the role of the commons. Let's look at a few laws from the third *Partida* regarding the commons.

Partida III, Titulo XXIX, Ley III: Quál cosas son que comunalmente pertenescen á todas las criaturas del mundo

What things that are communal that belong to every creature on

Earth. The things that belong communally to every creature that lives on this Earth are: the air, the rain from the sky, the sea and its riviera. Each creature that lives can use each of these things according to their needs. Each man can take advantage of the sea and its riviera for fishing, navigating and all the things that are understood to be part. But if in the riviera there should be a house or other building as long as it belongs to someone, it should not be destroyed nor used without the permission of its owner, unless it falls, then someone else can build another edifice in the same place.[11]

Ley VI: Como de los rios, et de los puertos, et de los caminos et de las riberas pueden usar todos los homes comunalmente

Like the rivers, the ports, the roads on the riviera, they can be used by all communally. The rivers, ports, public roads all belong to all men, the same they can be used by foreigners as well as those who live there. The rivieras along the rivers are based on the rights of those whose properties are nearby, but all men can use them as long as the boats are tied to the trees, with their sails, where they can put their merchandise, the fishermen can display their catch and sell it, all these can be used in the riviera as long as they are part of how they make their living.[12]

Ley IX: De quáles cosas pertenesce el señorio et el uso dellas comunalmente á todos los homes de alguna cibdat ó villa

What things belong to the territory and their communal use by all men in the city or villa. They are all communal in each city or villa, the fountains in the plazas where the merchants hold their fairs, the places where the council meets, the sand along the rivieras of the rivers, the other ejidos, where the horses have their races, the mountains and the grazing sites and all the other similar places that are established as communal in each city and villa. They are communal to all, the poor and the rich. Those that don't live there cannot use them unless given permission by those who live there.[13]

Ley X: Quáles cosas pertenescen á alguna cibdat, ó villa ó comun, et non puede cada uno dellos apartadament usar de ninguna dellas

> *What things belong to a city or villa as commons, and no one can use as their own.* Fields, vineyards, gardens, olive groves, and other properties, livestock, slaves and other things that produce an income, there can be cities and villas that are common to all who live there, but no one can use them individually for themselves. More over the income that is derived from any produce has to be used to benefit the whole community for roads, maintaining the castles, or those who take care of them, or anything that will benefit the community of the whole city or villa.[14]

What, then, are the common lands or ejidos? According to Page, "Since Roman times the *propios* had existed, which used to be rented for periods not exceeding five years. . . . As such in the *Siete Partidas* Alfonso X establishes that the cities can have *campos* (fields), *vides* (vineyards), *huertas* (gardens) and *arboledas* (groves), as well as pastures for the livestock that produce an income in order to benefit the public good, maintaining for example the walls and *portales* of the fortresses, sustaining the castles or making contributions (Partida III, Title XXVIII, Law X)."[15]

In the *Ordenanzas* of King Philip II the ejidos and *dehesas* are referenced explicitly in Ordinances 129, 130, 131, and 132.

Ordinance 129

> Within the town a commons (ejidos) shall be delimited, large enough that although the population may experience rapid expansion, there will always be sufficient space where the people may go to for recreation and take their cattle to pasture without them making any damage.[16]

Before a piece of land was applied for, the petitioners had to be very familiar with the area they wanted. To use a sports analogy, they had to scout the area and also follow the royal requirements, in this case the *Laws of the Indies* of 1681, which incorporated many of the 1573 ordinances. One of those requirements meant having sufficient public lands. As can be seen, the following is almost verbatim from Ordinance 129:

Book 4, Title VII, Law XIII:
Sufficient public land shall be designated for the town

> Public land shall be extensive enough that, if the settlement increases, there may always be enough space for the people to have recreation and for the livestock to graze, without causing damage.

Ordinance 130

> Adjoining the commons there shall be assigned pasture ground for the work oxen and for the horses as well as for the cattle for slaughter and for the usual number of cattle which the settlers must have according to these ordinances, and in a good number so they can be admitted to pasture in the public lands of the Council; and the rest shall be assigned as farm lands, which will be distributed by lottery in such a manner that the (farm lots) would be as many in number as the lots in the town; and if there should be irrigated lands, lots should be cast for them and they shall be distributed in the same proportion to the first settlers according to their lots; the rest shall remain for ourselves so that we may assign it to those who become settlers.[17]

This (Ordinance 130) became almost identical to what appears in Book 4, Title VII, Law 13:

> Pastures and lands shall be designated for public use. Having designated enough land for public use of the settlement and its growth, as has been ordered, those who have the authority to . . . establish a new settlement shall designate pastures that adjoin the public land for pasturing of work oxen, horses, livestock for butchering, and of the usual amount of other livestock that the settlers are ordered to have. They shall designate an additional proper amount of pastureland for the council. They shall designate the remaining lands as farmlands which they shall apportion by a drawing of chances, and there shall be as many of them as there are house-lots in the settlement. If there are irrigated lands, they shall likewise be apportioned in the same way to the first settlers by lottery, and the rest of these lands shall remain unassigned, so that we may grant them to those who come to settle later.

Ordinance 131

In the farmlands that may be distributed, the settlers should immediately plant the seeds they brought with them and those they may have obtained at the site; to this effect it is convenient that they go well provided; and in the pasture lands all the cattle they brought with them or gathered should be branded so that they may soon begin to breed and multiply.[18]

Ordinance 132

Having planted their seeds and made arrangements for the cattle in such numbers and with good diligence in order to obtain abundant food, the settlers shall begin with great care and efficiency to establish their houses and to build them with good foundations and walls; to this effect they shall go provided with molds or planks for building them; and all the other tools needed for building quickly and at small cost.[19]

Part of this legislation was also included in the *Recopilación* in Book 4, Title XVI, Laws II, IV, and VI. Also, Title XVII, Laws V, IX, and XII and Book 3, Title II, Law LXIII deal with the *Jueces de Agua*. In the *Novisima Recopilación* from 1788 under Charles II, Book 7, Title II, Law XXVII deals with acequias. The ejidos were also to be used for recreation and for livestock, since the dehesas were also included as part of the ejido. They were to be used for pasture for work oxen (*dehesa boyal*), for horses (*dehesa potril*), and for livestock for urban consumption or for meat markets (*carnicería*). Since the pasture was free, this lowered the price of meat, thereby benefiting the whole community. The word *dehesa* has its origin in Latin and signifies defense. In Spain after the reconquest, the Crown designated certain caballeros to defend the cities, giving each one a wide extension of land or dehesas granted under the generic name of *caballerías*, since they were assigned to gentlemen.

Titulo XVII, Ley VII:
Que los montes y pastos de las tierras de Señorio sean tambien comunes

That the mountains and pasture of the lands belonging to the territory also be common. The mountains, pasture, and the water within, and the mountains in the grants that are made, or are made from

> the territory in the Indies, they should be common to all Spanish and Indians. And thus we say to the Viceroy and the Tribunal, that they follow and obey.[20]

Ley IX: Que en quanto á los montes y pastos las Audiencias executen lo conveniente al gobierno

> *When it comes to the mountains and pasture the Tribunal should do what is best for the government.* The Viceroys and the Tribunal should do the best they can when it comes to pastures, waters, public houses and provide what is convenient to the population, and the perpetuation of the land. And they should provide written documentation of what is provided including the cost. And it is ordered, that justice be done to whomever inquires about the above.[21]

Ejido, then, is a Latin word that means *exitus*, or exit, outside the city signifying a collective space, which could be used for a "common pasture." These communal lands that belonged to the *baldíos* or *tierras realengas* included the dehesas and *propios* among other names such as *cotos*, *prados*, *entrepanes*, *tierras arables comunes*, *entradizas*, and *cañaderas*. According to the legislation, they were supposed to be different in practice and among the people were defined as ejidos. The more common names were *dehesas*, *propios*, and *baldíos*. The dehesas were also part of the ejidos. Baldíos were lands that were not in use and belonged to the Crown, and they were also called *realengas*. They were usually large pieces of land that the Crown permitted to be used at its discretion, mostly for common pasture.

And though each vezino (settler) had individual suertes (long lots below the acequia), the dehesas (commons) belonged to everyone.

Titulo XVII, Ley V: Que los pastos, montes, aguas y terminos sean comunes, y lo que se ha de guardar en la Isla Española

> *That grazing, mountains, water, and boundaries be in common.* It is ordered that the use of all grazing, mountains, and waters in the Provinces of the Indies be common to all the settlers, who are now settled, or will settle, so that they can enjoy freely . . .[22]

Not only were all pastures, mountains, and water to be shared by all, but once the harvest was in, the stubble fields became common lands.

Ley VI: Que las tierras sembradas, alzado el pan, sirvan de pasto comun

> *That the land planted in grains, once harvested, they serve as common grazing.* The lands and property within the grant, or bought in the Indies, once harvested, remain common pasture lands.[23]

This was still practiced throughout northern New Mexico and in the Embudo grant as late as the mid-1950s. Even into the 1970s, some people would cut the fences to allow their cattle onto the *rastrojos*. The land, which had the rastrojos in a way, was both private and communal, which made it a unique concept in land ownership. It was usually part of the suerte, the long lot, which made it private property, but once it was harvested, then it became communal for grazing purposes.

Also common were the fruits grown in the wild.

Ley VIII: Que los montes de fruta sean comunes

> *That the fruits in the mountain be common.* Our desire is to make all wild fruits in the mountains common, and that each individual can harvest and take plants to transplant in their property or farm, and advantage should be taken being that they are common.[24]

Even today people go and harvest plants and trees from Forest Service and BLM lands because they consider them part of their heritage, and they don't see themselves breaking the law, because of this particular law.

The laws already cited, for this particular essay, are what this writer considers the most important in attempting to understand how a land grant was made and what had to be done to continue in possession of the land. Contrary to popular belief, the vezinos had to prove that the land they desired was vacant, that it didn't infringe on the Indians who lived close by, and that the land was fertile and had good climate. It also required them to observe the Natives to see if they lived to old age and if they were healthy; they were also concerned with air and water pollution. Their houses were to be built on the lands above the irrigated fields.

Ricardo Legorreta, a Mexican architect, wrote in the *New York Times*, "You cannot deny your parents. You cannot deny your history, your roots."[25]

For material objects were not the only items that made their way from Spain to Mexico and then to New Mexico. Probably the most important immaterial entities that traveled from the South to the North were the ideas and philosophies as to how people related to land and water use in a new environment very similar to that of the Iberian Peninsula. To understand these ideas one has to unravel the *trenza*, or braid, one strand at a time; but for this system to work the strands have to be braided together.

Wrote Akihiro Kashima, Takashi Tanimura, Tsuyoshi Kigawa, and Masao Furuyama, in *Influence of the Italian Renaissance on the Town Planning Concept in the Spanish Colonial Laws*, "Many of town planning experiments in the Spanish Renaissance period were carried out along with the new town settlements under the oversea colonial administration. The town planning concepts illustrated there have been attracting attention not only for its [*sic*] practicality but also for its correlation to the ideal cities in the Italian Renaissance. Parallel to the practice of town planning, the Spanish Royal issued many colonial laws, *Leyes de las indias*, and provided regulations on town planning. The ideas in those regulations attract much attention for its influence on the modern town planning concepts also."[26]

When Legorreta talks about not denying your parents, history, and roots he is referring to the Arab influence that reaches back to the Spanish colonizers' Moorish past and also our Roman roots. In trying to understand the so-called New World, it is usually a black and white dichotomy of Spanish (Castilian) versus Indigenous, i.e., Mesoamerican, influence. And even here, Mayas and Aztecs take most of the credit while the Tlaxcaltecas who came with the early settlers under the Spanish Crown are not even mentioned though they settled early on in Santa Fe, probably as early as 1600, and also in Albuquerque, at Atlisco as in the Atrisco land grant. But history is not that simple—there are lots of shades of gray in the palette. It is only recently that the Sephardic, or Crypto-Jewish, tradition has begun to be studied. What for all practical purposes is not even mentioned in scholarship is the Muslim influence, though about a third of all Spanish words are derived from the Arabs. When analyzing the land patterns in New Mexico we always go back to the *Recopilación de las leyes de los reinos de las indias*, known as the *Laws of the Indies* of 1681, which are based on the *Ordenanzas* of King Philip II of 1573. But like peeling an onion, when you start searching for the laws' antecedents, we encounter the Arab influence in all aspects of land and water use in New Mexico, albeit under the guise of Roman law. The Moors are the prodigal sons.

Under the *Laws of the Indies*, the land was divided into what we know today simply as commons and the irrigated lands. What divides the one from the other is a rigid zigzag line formed by the acequia, the channel that delivers the water and gives life to all the land below it. This rigid design line follows the contours of the land. Above the acequia is the dry land, which is more in tune with how the land was managed in northern Europe prior to the arrival of the Arabs on the Iberian Peninsula in 711. When the Moors were kicked out of Spain, how they managed the land did not disappear. In fact, it resurfaced in the "Indies" under the guise of different ordenanzas, the laws under which the Spanish land grants were made to settlers.

Acequias, as a landscape, are a combination of private and common lands tied together by the water that nourishes both, which flows from the watershed usually born high up in the sierras. A watershed can be part of a small tributary like the Río Embudo, which is made up of several creeks that eventually flow into the Río Grande. Or acequias can be directly diverted from the main trunk as is the case with the acequias in Velarde, which all draw their water from the Río Grande. Acequias are usually always born in the commons since the *toma*, their place of birth, and the *presa*, the structure that diverts the water, are in the river where they are born. Their *desague* is again in the commons since they empty back again into the river, usually the river in which they are born, although not always.

The Acequia Junta y Ciénaga, which draws its water from the Río Embudo, empties into the Río Grande about half a mile from the "junta de los ríos"; that is, where the Río Embudo empties into the Río Grande. The Río Embudo is born high up on the western slope of the Sangre de Cristo Mountains, which are part of the southern Rockies. Boundaries of the watershed, which covers about three hundred square miles, are in the form of a triangle and thus the name *Embudo* or *funnel*. It is bound on the north side by the Cañon de la Junta, which is its right arm (looking west); the center of the watershed is the majestic bald mountain, known as La Jicarita. Its southern arm is made up of the mountain range that includes La Jicarilla Peak, El Chimayoso, and the north side of the second highest peak in New Mexico, the Truchas Peak. The water from these three peaks forms the Laguna Escondida (Hidden Lake) and Laguna de Abajo (Lower Lake), where the Río de Trampas is born. The Río de Santa Bárbara is born at the bottom of the Jicarita, where the water from the west face of the Jicarita comes together. Another small rivulet is born on the southwest side of the sierra, known as El Río Chiquito, which merges with the Santa Bárbara near Peñasco. Meanwhile, the Río del Pueblo, which is the third creek that later comes together to create the Río Embudo, is born on the north side of the watershed, or Cañon de la Junta, and Ríto de los Alamitos.

This is a description of the Embudo Watershed. In Spanish, a watershed is known as a *cuenca* (*territorio cuyas aguas afluyen todas a un mismo río, lago o mar*), a territory whose waters flow to the same river, lake, or sea. This landscape, which resembles the head and shoulders of the human body, is all common land managed by the Forest Service as part of several land grants that date back to the 1700s and, prior to that, Native American lands that belonged to the Picurís.

To the Moors who settled southern Spain, more commonly known as *al-Andalus* or *Andalucía*, the communal lands were known as *harim* and *mubaha* and to the Arabs as *mawat*. In Spanish these became known as *tierras muertas*, or dead lands. These were lands that could not be sold and were to be used in common by all the people of the land grants, or *alquerías*, as these lands were known in Andalucía. *Alquería* is derived from *alcarria* (*qarya*) and signifies *aldea*, a village or hamlet. *Aclaria* is a lady's name in New Mexico. Dr. Carmen Trillo San José from the Universidad de Granada recently wrote an excellent book on this topic, *Agua, Tierra y Hombres en Al-Andalus: La Dimensión Agrícola del Mundo Nazarí*. Land grants were usually composed of more than one aldea or hamlet. In the Embudo land grant there are twelve separate hamlets, and all the historic names refer to the landscape, i.e., *Cañoncito*, *Montecito*, *Apodaca*, *Bosque*, *Angostura*, *Junta*, *Ciénaga*, *Nasa*, *Rincón*, *Bolsa*, *Rinconada*, and *Plaza del Embudo* (today Dixon).

Many people in the Río Arriba Bioregion, when referring to the commons, think of ejidos, which simply supplanted the word *latifundia* here. And though the term *land grant* has a high recognition level among the general population, especially the Indo-Hispanos, very few understand its anatomy. Latifundias are big expanses of land, in the thousands of acres, whereas minifundias are small land holdings of only a few acres. And *ejido* simply means *exitus*, or the place at the outskirts of a village, which is neither planted nor worked and is common to all. It's from the Latin verb *exeo, exis*—to exit, to leave. There are four main divisions within an ejido, or the commons, even though they blend and overlap into each other at times, again, like a braid:

Sierras
Montes
Dehesas
Solares

Sierras provided the early settlers—and today's descendants of these early pobladores—a place to harvest firewood and also *vigas* and *latillas* for constructing houses and other buildings needed for survival. When the mercedes or land grants were awarded, building materials for living quarters were dragged from the sierra and *monte* using animal power; today trucks are employed for this type of labor. The settlers also combed the lands for wild fruits, *capulín* (chokecherries), *chatacow* (elderberries), *moras silvestres de matas y de suelo* (wild raspberries, alpine strawberries), piñon, and *beyotas* (acorns). Wild herbs, such as *oshá*, *oregano de la sierra y del campo*, *altamisa*, *poleo*, and *yerbabuena* were and are harvested today. Each village has their places where certain essential herbs are grown and harvested; many of these sites are kept secret. Since the coming of the flower children or hippies, many of these sites have been raided to the point of near extinction as some started harvesting the herbs to sell commercially.

Curanderas, who in the past readily told others of where they got their *remedios* or herbs, today are cautious as to whom they divulge their secret gathering places, whether in the high sierras, the juniper and piñon montes, or the high-desert dehesas that produce the *chimajá* (or wild parsley) in the spring. Those familiar with the language know that most landscapes are named to signify where certain raw materials are found. For example, *el llanito del zacate de la escoba* meant broom grass grew there, while *el arroyo del barro* identified the site as a clay deposit. Place names also related to the local environment. *Costilla*, for example, meant the mountains looked like ribs, and *Questa* signified going up or down the side of the mountain or costilla. *Embudo*, as discussed, means *funnel*, and the Embudo Watershed is in the form of a funnel.

Like the allocated lands, these communal lands or ejidos were broken down into sierras, montes, dehesas, and *solares* where the houses were built. But the commons were also crisscrossed by *cañadas* and *veredas*. A cañada can be described as a *camino mesteño*, a wild road, since they were used to move the livestock, mostly sheep and goats, from the winter to summer pastures and vice versa, from the dehesas to the sierras. A cañada is usually defined as a space between two high peaks (*lomas*) or mountain ridges (*cuchillas*) that has water holes, or *abrevaderos*; includes vegetation for animals to eat; is at least ninety *varas* (a little less than a yard—around thirty-three inches) wide; and is mainly used to move livestock. Besides abrevaderos, cañadas also have spaces where the livestock rest called *descansaderos* or *majaderos*, which refer not only to a resting place but also where manure is deposited. Also part of the commons are the veredas, or trails, which are more narrow but usually a minimum of twenty-five varas and usually used by horses or to move smaller flocks or herds of livestock.

There is a *dicho*, or saying, that says, "Quien deja el camino real por la vereda, piensa atajar y rodea [He who leaves the royal road for the trail thinks he will make a shortcut but instead makes the road longer]." Both cañadas and veredas are common roads. It's from the *cañadas reales* that the term *dehesa* might have originated, according to some scholars, since this caused a conflict between those moving livestock and the inhabitants of the villages through which the animals were moved twice a year. From there, the term *defendere*, which means *permission*, *dehesa* is thought to have originated, since the king had to intercede and grant permission. All of these concepts eventually made their way into the *Laws of the Indies* and thus to New Mexico.

Sierra defines a mountainous terrain whose features resemble the teeth of a saw. The term seems to have its roots in the Sabaeno *as-sirr*, which refers to a rugged high desert, from fifth-century BC southern Yemen. In Spain the word applies to high, saw-tooth mountains, and it was appropriately transferred to the southwestern ranges by the settlers under the Spanish Crown. It's in the sierra where the cuencas (watersheds) form, and they act as the keepers of the water because the snow melts slowly, thus providing not only the irrigation water for the acequias but also feeding the aquifers that feed the *norias* (another Sabaeno word, from *an-naura*—the same word is used in Syria) or wells that provide the water for domestic uses.

Monte is derived from the Latin, *mons—tis tierra alta*, high ground—while *montaña* is *tierra alta, áspera y habitata*; that is, highlands, harsh but habitable. There is a verse that says the following:

Preñado dicen que estoy	Pregnant they say I am
Y jamas a parir vengo,	But I come not to give birth,
Lomos y cabeza tengo	Loin and a head I have
Y aunque vestido no estoy	And though dressed I am not,
Muy grandes faldas mantengo.	Huge skirts I have.

It is said that the mountains are pregnant because of their huge "rumores y hinchazones." They appear pregnant because of their swellings and bulges. *Tienen tambien cabeza y es su cima y espaldas y sus vertientes llamamos faldas, aunque no ande vestido, y dicen comunmente falda de un monte*; they also have a head (the summit) and also shoulders (the slope of the watershed). Mountains also have *cejas*, or eyebrows. The English language is not that precise when it comes to naming the environment, since personal names are more commonly used. An example is the Pedernal Peak near Abiquiu that people wanted to rename O'Keefe Peak for the renowned artist Georgia O'Keefe, but she had more sense and declined the renaming before she died. *Pedernal* means *flint*, reflecting the fact that there's a lot of flint in the area.

The nonirrigated lands of the mercedes, especially those lands known as *secano*, used for dry farming, are usually on the lower reaches of the dehesas, known as *tierras de pasteo*, or pasture lands. In Latin the dehesa is called *pascua*, and it is a place where the livestock graze. It could very well come from the Roman custom of establishing latifundias in marginal lands. But the term does not appear until the year 924 in the Corominas dictionary, though the term is also found in the laws of the Visigoth, known as *pratum defensum*, as noted by the Romans. According to Covarrubias, it is an Arab term that means "a low land, full of weeds where it is hard to walk, from the moisture in the soil and thick with weeds." Covarrubias says the word comes from *dehisetum*, from the verb *dehesa*, "que vale espesar y estrechar." But he says it could also be Jewish, from *dese*, *herba*, for the *deshesa* is nothing more than "a piece of land full of weeds." A dehesa is a seminatural ecosystem where there is usually a certain amount of human involvement. In New Mexico this meant that the piñon trees were pruned to the extent of removing what is known as *piñon blanco*, the dead piñon branches that have gotten a gray patina and were treasured by the ladies when they relied on firewood for cooking and heating, for it is seasoned wood. Also this type of piñon tree is the one that usually produces the best piñon nuts, and because it has been taken care of, the nuts are easier to harvest.

A dehesa is also a space that conserves a great number of both flora and fauna; it also has great economic and social importance. Regardless of its original meaning, whether *dehesa* has Latin, Arabic, or Hebrew roots, it is understood to be an agro-forestry system with poor soil and a harsh climate where man has intervened to make it somewhat productive. Some scholars say that a dehesa is not very ecological due to the economic pressures of grazing more livestock than it can sustain. Dehesas once formed part of the different land grants and are now managed by the Bureau of Land Management, the State Land Office, and the Forest Service. It is usually a type of pasture with scattered trees of evergreen piñon and juniper (*cedro y sabina*) and deciduous oak, and in the past grains were often grown under the sparse tree covers. The space then between the dehesa and the solar is what was used for dry farming. It was situated above the rigid line made by the acequia, which separated the commons from the private lands.

A dehesa can be better understood as a mosaic because of its different uses; it's also part monte but is also used for grazing and, when necessary, dry farming. The best pinto and *bolita* beans are grown on secano. It's an agroforestry system with the joint production of trees and agricultural crops and/or animals; it's also known as an agro-silvo-pastoral system.

Besides the sierras, montes, and dehesas, though private and to a certain extent part of suertes, and usually above the acequia, are the solares where the houses were commonly built. *Solar* (roughly 130 feet by 130 feet or 50 varas by 50 varas) comes from the word *suelo*, to make a floor as in constructing a house on a plot of ground. But a solar is not only the site where the house is built, it also is the space between the acequia and the commons where the settlers built their corrals, *gallineros*, *trochiles*, and *leñas*; that is, the space where the corrals, chicken coops, pig pens, and wood piles were kept. The house, if away from the town plaza, was constructed following an *L* shape or *U* shape, the same as the Moorish houses on the alquerias. Also, part of the house complex included the *dispensa*, or utility room, and *soterrano*, or root cellar, where people kept their food supplies for winter.

Rome conquered Spain in the fourth century before the birth of Christ and had a vast influence on everything, including agriculture, especially when it came to land use. They introduced the aqueducts that were able to transport water to areas that were without water before, and though Roman impact was more visible in urban areas, their introduction of the plow had an impact on agriculture as well.

La Acequia: The Lifeblood of Querencia

What I call agricultura mixta tradicional mestiza, which I have been practicing for more than twenty years, is based on irrigation, dry farming, and natural farming. The most important element in our three-pronged agriculture is the acequia.

According to the ordinances, first on their agenda was to lay out the acequia madre, as water was needed for everything. The acequias were an artificial system for moving the water from the river, or in some cases springs; then once constructed, planting began, as was the law. And even before building a house, they should "plant trees," as recommended by Gabriel Alonso de Herrera in his *Obra de agricultura* of 1513 and also in the *Laws of the Indies*, as stated earlier.

Ley Xj: Que las tierras se rieguen conforme esta ley

> *That the lands be irrigated according to this law.* It is ordered that the same order that the Indians had in the division and allocation of water be observed and practiced among the Spaniards where it has to be divided, and the land being marked, and for this involve the natives, that used to be in their charge, that they be watered the same, and each one gets the water, which they are entitled, successively from one to the other, with penalty for those who want priority and take it, taking the law onto their own hands, it should be taken away from them, until everyone has watered their field.[27]

In *Riegos en la Nueva Vizcaya* by engineer Victor Mendoza Magallanes, he cites Partida III, dealing with acequia easements: "That such easement shall be twice as wide as the bottom of the acequia, or four steps of Salomon, measured on each side of the acequia, of which such easement no one shall claim because it is community property."

Libro 3, Titulo II, Ley Lxiij:
Que dá la forma de nombrar Jueces de aguas, y execucion de sus sentencias

> *How to appoint water judges and how they should carry their duties.* It is ordered that the Tribunals appoint Judges, if it is not the custom, then the Viceroy, or President of the City or Town, that distribute the water to the Indians, so that they can irrigate their farms, gardens or seedlings, and water for the livestock as long as they don't do damage and divide those they have. Once the division is complete, let the Viceroy or President know how they should proceed. And we order that the Indians should not be charged, and in things that they know, if their judgments are appealed, it should be what the Tribunal determines.[28]

Though the law is very clear that the pobladores follow the Natives' tradition and custom by including it in the law—"It is ordered that the same order that the Indians had in the division and allocation of waters be observed and practiced"—toward the end of the same law it invokes the custom and traditions brought by the Moors to al-Andalus in southern Spain: "And each one gets the water, which are entitled successively from one to the other, with penalty for those who want priority and take it, taking the law onto their own hands." In essence this law takes into account the customs and traditions of both Indians and Moors and creates a new set of laws that is truly New World. The laws in Spain prior to 1492 regarding water were almost identical, in that Sarracenos, which referred to the Arabs, substituted the word *Indio*.

Next, once they had dug the acequia and planting was started, they had to start building their houses.

> *Ley XV: Que haviendo sembrado, los pobladores, comiencen á edificar*
>
> *That having planted, the settlers should start building.* Once the planting has been done, and the livestock is in place, and with the help of God a good crop is expected, begin with a lot of care and diligence to build your houses with good foundations and walls, and start constructing retaining walls, garden plots, and have all tools and instruments needed for building fast and economical.[29]

> *Libro 6, Titulo VII, Ley XXVI: Que los pobladores siembren luego, y echen sus ganados en las dehesas, donde no hagan daño a los Indios*
>
> *That the settlers plant, then take the livestock to the common lands, where they won't cause any damage to the Indians.* Then, and without delay, that the agricultural lands be divided, the settlers plant all the seeds that they had with them, and they should be prepared; and for greater facility that the Governor appoint a person to be in charge of planting the grains and vegetables to better supply themselves; and in the common lands keep all the livestock that is possible, with their brands and marks, so that they can start breeding and multiply, in places where they will be safe and won't cause any damage in the lands, planted fields or any other property belonging to the Indians.[30]

As can be seen from these laws, as the settlers were cautioned by the governor when the Picurís Indians raised their concern about the livestock of the vezinos damaging their fields, what the governor referred to was the law just cited.

Even to build a house, the settlers had to follow the royal decrees. More than likely, for the first few years the pobladores lived in *jacales* (a temporary house made of sticks and mud), though according to royal requirements, all houses had to conform to a certain style, three hundred years before Santa Fe Style was mandated by law.

Ley XVII: Que las casas se dispongan conforme a esta ley

> *That the houses be built according to his law.* The settlers agree, that the lots, building and houses follow a certain pattern, with the same adornments, and that they enjoy the north and south winds, uniting them, so that they can serve as a defense and fortification for those who would want to obstruct or infest and by all means all houses should have enough space for the horses and beasts of burden, with courtyards and pens as wide as possible, so that they be healthy and clean.[31]

According to such an ordinance, all houses had to be united in the form of a *plazuela* to better "serve as a defense and fortification." All while building and working, the settlers were to avoid contact with the Natives, though this might have been easier mandated than done. Whether that happened in practice is not known.

Following the *encuentro* of the two hemispheres, the way the land was worked changed immensely, mostly due to the introduction of new techniques and tools. Possibly what made agriculture flourish, especially in the arid landscapes such as northern Mexico and New Mexico, was the introduction of the *arado*, or plow. Before, the indigenous cultures relied mostly on the use of wooden tools, such as the *coa*, similar to our *cavador* or hoe. But the plow allowed the farmer to open up the soil, to loosen it up and turn it over, and then to deposit the seeds so they could grow.

The plow and hoe haven't always gotten along, as can be seen in a very early *trovo*, a type of poetry that came from the poor who labored the soil, whose origins go back to North Africa. Titled "The Disputation Between the Hoe and the Plow," it comes to us from biblical times and has long been recognized as one of the first poetic statements of the common man against the rich and mighty. There are a total of twenty-six stanzas; here are a couple of them:

> Hey! Hoe, Hoe, Hoe, tied up with string;
> Hoe, made from poplar, with a tooth of ash
> Hoe, made from tamarisk, with a tooth of sea-thorn;
> Hoe, double-toothed, four-toothed;

Hoe, child of the poor, bereft even of loincloth;

O Plow, you draw furrows—what is your furrowing to me?
You make clods—what is your clod making to me?
You cannot dam up water when it escapes,
You cannot heap up earth in the basket,
You cannot press clay or make bricks
You cannot lay foundations or build a house
You cannot strengthen an old wall's base
You cannot put a roof on a man's house
O Plow, you cannot straighten a street.
O Plow, you draw furrows—what is your furrowing to me?
You make clods—what is your clod making to me?"[32]

The Berbers also took this type of poetry to the Alpujarras south of Granada, and it eventually made its way to Mexico and to New Mexico. It was not limited to tools of the trade and techniques, since there was a big difference between the hoe and the plow. Here in New Mexico it surfaced as the *Trovo del Café y el Atole*, where *Café*, or coffee, represented the moneyed interests and *Atole*, corn gruel, the indigenous poor farmer. As with the plow and the hoe, in this trovo, Atole makes Café succumb also. Here is also a sampling:

Por mi gracia y por mi nombre
By my grace and by my name
Yo me llamo don Café.
I am called Mr. Coffee.
En las tiendas más hermosas
In the most luxurious stores
Allí me hallará usted.
There you will find me.
A la América he venido
To America I have come
Y es claro y evidente
And it is clear and evident
Desde mi país he venido
From my land I have come
A conquistar a tu gente.
To conquer your people.

Verdad yo soy el Atole
True I am Atole
Y a Dios le pido la paz.
And God I ask for peace.
Café que recio vas.
Coffee don't go so fast.
También yo te dire
I will also tell you
Que muchos en el estribo
That many in the stirrup
Se suelen quedar apie.
End up having to walk.

As can be seen from the examples of the two different trovos, one from biblical times and the other hundreds of years later and in a different continent, though both deal with working the land, with techniques and crops, they are almost identical in that both represent the struggle between the poor and the mighty, and in both cases the underdog comes out on top. So not only did agriculture cross-pollinate in terms of techniques but also in terms of crops. Today both coffee and corn are money crops that have heavily influenced agriculture. Coffee was supposedly discovered by an Arab goatherd who observed his flock eating the coffee beans and realized that after the goats ate the beans it changed their behavior. Corn, on the other hand, was a crop that one can say was developed and nurtured by man for thousands of years. Where coffee might survive on its own in the wild, corn cannot survive without the hand of man.

Therefore we can see that agricultural techniques and crops have also influenced the poetry of the common man, from North Africa to the Iberian Peninsula, then to Mexico, and eventually to New Mexico.

For if it hadn't been for the plow, first the wooden plow and finally the iron plow, agriculture wouldn't have been able to spread and grow the way it did. And with the invention of the plow also came the domestication of the beasts of burden to work the plow. First it was the ox, and then the mule and the horse that were used to work the land. Prior to the arrival of the ox, mule, and horse, the indigenous populations didn't have animals that were domesticated to help them do agricultural work on a massive scale.

Though some of the techniques are touted as innovative by the current organic and sustainable agricultural movements, they have been done by the indigenous people for hundreds of years. Double digging as promoted by the biodynamic practitioners is nothing new to the *chinamperos* of Xochimilco. While in Xochimilco doing research on the chinampas, I noticed a campesino that had never heard of double digging, turning the black organic soil with the shovel to a depth of about twenty-four inches.

It is here then that the farmer will transplant the tiny plants he grows individually in his *chapínes*, also known more commonly as *almácigos*, or plant nurseries. In Latin this idea is known as *atajo de tierra*, for it is essentially small beds where plants are grown from seed. The word *almáciga*, or *almácigo*, is also common in northern New Mexico among the older traditional farmers. It comes from the Arab *al-maskaba*, which means an irrigated piece of land. What we see is that both hemispheres had similar techniques already in place, whether it was called a chapín or almáciga, and after five hundred years the words are used interchangeably and understood by both the indigenous as well as the mestizo farmers. I remember that my mother always prepared her almácigos, starting around St. Patrick's Day, or March 17, so that they would be ready to transplant by early May.

But it is not only in the techniques of preparing the soil and plants where we find the Arab influence, but also in how the land was divided and appropriated. That is, the mercedes or grants of land were very similar to the Arab alquerías. And where a merced is composed of both irrigated and nonirrigated lands, we find the same type of land divisions with the Arabs. What we call *ejidos*, or the common lands composed of sierras, montes, and dehesas (pasture lands), to the Arabs were known as *mamluka*, or appropriated lands, which would be similar to our suertes or irrigated pieces of land. The nonappropriated lands were known as *mubaha*, and it is these lands that people could use for pasture of their livestock or flocks or to go for wood or wild fruits and plants. These lands were known as communal or *harim*; then there are those known as tierras muertas or mawat that appear to be more like solares since they were used for houses. These can be acquired by simply living on them, but none could be alienated or sold. If left vacant for two years they could lose the use of the land. This was similar to the land grants, but in land grants the time period was four years. Like the land grants, this type of land division permitted both livestock production and intensive agriculture, this in the irrigated portions of the land.

What provided the intensive agriculture then was the acequia system, which was an elaborate and complex system of managing the water. The same as in contemporary New Mexico, the water was diverted from the river, starting at the toma, or where the water was taken by *presa* or *azud*, and from there it was moved by the principal canal known as the *acequia madre*, or mother ditch. From there it was conducted to the planted fields via *acequias secundarias* or *menors*, also known as lateral ditches or *linderos*. These laterals then were diverted into *hijuelas*, which run parallel, and *cabeceras*, which are horizontal ditches, and from there to the *brazos* and eventually to *ramos* to irrigate the *bancales*, *bancos*, *ancones*, and *azoteas*—four different types of terraces. The *bancales* were terraces on slopes, whereas the *bancos* were bigger and usually in valleys, and the *ancones* were the small terraces or coves along the riverbeds. I have identified a fourth type of terrace, called *azotea*, from *flat roof*. Then to irrigate the terraces, the *regadera* or *compuerta* would be opened to let the water in or closed to stop the flow of water. Then the *escurriduras* would be picked up at the end of the furrows, also known as *carreritas* or *surcos*, and from there combined with other excess water known as *azarbeta*, *azarbeta menor*, and *azarbeta mayor* in Murcia, and here simply as *desagues*, and returned to the river. The *escurriduras* were common property, since once the farmer is finished irrigating, the excess water is allowed to flow back to the river or to another piece of land.

Suertes (two hundred varas [each vara is approximately thirty-three inches] across by four hundred in length, or approximately thirteen acres) were then broken down into *altitos*, or the highest terrace right below the acequia. Then came the *joya-jolla*, or jewel-hallow, the most fertile piece of land, followed by the *vega*—where most people kept their domestic animals but it could also be used as a corn patch or milpa—and finally came the *ciénaga* or wetland. A ciénaga can also be used for growing crops if it is drained, or *sangrada*, similar to when one is injured and blood has to be bled to relieve the pain and pressure.

In New Mexico when the water is diverted from a *cabecera*, it is also known as *sangria*. In San Miguel Tlaxipan also near Texcoco, these land divisions are known by the size of the plants or trees growing such as *arboles grandes*, like *aguacate* or avocado, and then *arboles medianos*, such as pears and apples. The third land division is where *plantas arbustas* such as *romero* and *ruda* are grown. Then comes where the flowers grow, *plantas de flores*, and finally, *plantas rastreras*, such as *hierba buena* (mint), is grown.

The same type of agriculture that we have in these types of land divisions, whether in New Mexico, Mexico, Spain, or the Middle East, is watered by the acequias, known in Ladakh, India, as *yuras*. The presa is known as a *raks*. The mayordomo (known as a *Juez de Agua* in the Ordinances) in New Mexico is known as the *cequier* in Valencia and *aguador* in Chihuahua. In Ladakh he is the *chud-pon*—different name but the same responsibility. *Chud-pon* is derived from *chu*, meaning *water*, and *pon*, the appropriator of water. Then we have natural agriculture, known in New Mexico as the jardin *rizo* or *ricio*, those plants that volunteer on a yearly basis without the benefit of having to plant them. Here people gather the wild asparagus, purslane (*verdolaga*), plants not native to the New World but that have become accustomed here and now grow wild. They also gather *quelites*, a relative to wild amaranth and quinoa, known as *quelite juz* or *quelite del burro*, *quintonil* in Mexico, *quelite pardo* in New Mexico, and *quelite cenizo* in the chinampas.

Conclusion

What I have written here is a very brief outline of where our traditional agricultural roots come from, the written ordinances dating back to 1265. As Nuevomexicanos we cannot neglect one aspect, whether the European, African, or Mesoamerican roots, if we want to fully understand traditional agriculture as practiced by the Indo-Hispanos in la Nueva México.

We use the Roman plow to open the land, to plant our American chile seeds following the chinampas tradition of intensive planting, while watering using the Moorish acequia tradition. For example, if you visit a traditional northern New Mexico home during the lunch hour, or for supper, you might be served red chile with pork, fried potatoes and corn, fabas, and wheat tortillas. And for dessert, flan, introduced by the French, or *arroz con leche y pasas* (sweet rice with milk and raisins, an Arab dessert) and *cuajada* (a sort of yogurt). The origins of these foods are from throughout the globe.

We owe a debt of gratitude to Columela, Zacarias, Herrera, and the countless Mesoamerican agriculturalists, whose methods of traditional agriculture all ended up in northern New Mexico. And here they were not only preserved but also refined by people such as my parents, and we still practice and maintain today these old permacultural practices. As *labradores*, remember "that we do not inherit the land from our parents, we have it borrowed from our kids," and for that reason each generation has to improve on what the last has done. When all is seen together, what we have is that which defines the querencia of the Nuevomexicano of the Río Arriba Bioregion, from the language, cuisine, poetry, and agricultural practices that give a certain perspective on managing the commons. But commons don't only refer to land but also to water and folklore, and all are there to be used by the common man, rich or poor.

NOTES

1. **34.** Para hauer de poblar, así lo questá descubierto, paçífico y debaxo de nuestra obediencia como en lo que por tiempo se descubriere y paçificare, se guarde el orden siguiente: elíjasse la prouincia, comarca y tierra que se a de poblar teniendo consideraçión a que sean saludables, lo qual se conocerá en la copia que huuiere de ombres viejos y moços de buena complisión dispusición y color y sin enfermedades y, en la copia de animales sanos y de competente tamaño, y de sanos frutos y mantenimientos, que no se críen cossas ponçoñossas y noçibas, de buena y felice costelaçión, el çielo claro y begnino el ayre y suaue sin impedimento ni alteraciones, y de buen temple, sin excesso de calor o frío; y hauiendo de declinar, es mejor que sea frío.
2. **35.** Y que sean fértiles y abundantes de todos frutos y mantenimientos y de buenas tierras para sembrarlos y cogerlos, y de pasto para criar ganados, de montes y arboledas para leña y materiales de cassas y edificios, de muchas y buenas aguas para bever y para regadíos.
3. **39.** Los sytios y plantas de los pueblos se elijan en parte adonde tengan el agua çerca y que se pueda deribar para mejor se aprouechar della en el pueblo y heredades çerca del y que tenga çerca los materiales que son menester para los ediffiçios y las tierras que han de labrar y cultibar y las que se an de pastar, para que se escusse el mucho trabajo y costa que en qualquiera destas cosas se habrá de poner estando lexos.
4. **40.** No se elijan en lugares muy altos, porque son molestados de los vientos y es difiçultoso el seruicio y acarreto, ni en lugares muy baxos, porque suelen ser enfermos; elijan en lugares medianamente lebantados que gozen de los ayres libres y espeçialmente de los del norte y del mediodía; y si ouieren de tener sierras o cuestas, sean por la parte del poniente y de lebante; y si por alguna caussa se ouieren de edificar en lugares altos, sea em parte adonde no están sujectos a nieblas haziendo oserbaçión de los lugares y açidentes y hauiéndose de edificar en la ribera de qualquier río sea de la parte del oriente, de manera que en saliendo el sol dé primero en el pueblo que en el agua.
5. *Libro 4, Titulo VII, Ley III. Que el terreno y cercancia sea abundante y sano.* Ordenamos, Que el terreno y cercancia, que se ha de poblar, se elija en todo lo posible el mas fertil, abundante de pastos, leña, madera, materiales, aguas dulces, gente natural, acarreos, entrada y salida, y que no tengan cerca lagunas, ni pantanos en que se crien animales venenosos, ni haya corrupcion de ayres, ni aguas.
6. *Libro 4, Titulo, XII, Ley XI. Que se tome possesion de las tierras repartidas, dentro de tres meses, y hagan plantios, pena de perderlas.* Todos los vezinos y moradores á quien se hiziere repartimiento de tierras, sean obligados dentro de tres meses, que les fueren señalados, á tomar la possession de ellas, y plantar todas las lindes, y confines, que con las otras tierras tuvieren de sauces, y arboles, siendo en tiempo, por manera, que demás de poner la tierra en buena, y apacible disposicion, sea parte para aprovecharle de la leña, a que hubiera menester, pena de que passado el termino, si no tuvieran puestas las dichas plantas, pierdan la tierra, para que se

pueda proveer, y dar á otro qualquiera poblador, lo qual no solamente haya lugar en las tierras, sino en los Pueblos, y canjas, que tuvieren, y huviere en los limites de cada Ciudad ó Villa.

7. Instructions given to don Pedro de Peralta when he was named governor and captain general of New Mexico, March 30, 1609, México. This came from some documents given to the author by Santa Fe historian José Garcia, during Santa Fe's Cuarto Centenario in 2010.
8. Kirkpatrick Sale, *Dwellers in the Land: The Bioregional Vision* (San Francisco: Sierra Club Books, 1985), 42.
9. Sebastiao Salgado, introduction to *Migrations: Humanity in Transition* (New York: Aperture, 2000).
10. Carlos A. Page, "Los Ejidos como Espacio Comunal de la Ciuda de Córdoba del Tucumán," *Revista de Indias*, vol. LXIV, núm. 232 (Argentina: Universidad Nacional de Córdoba, 2004), 635–50.
11. *Partida III, Titulo XXVIII, Ley III Quál cosas son que comunalmente pertenescen á todas las criaturas del mundo.* Las cosas que comunalmente pertenescen a todas las criaturas que viven en este mundo son estas: el ayre, el las aguas de ls lluvia, et el mar et su ribera; ca qualquier criatura que viva puede usar de cada una destas cosas segunt que fuere meester: et por ende todo home se puede aprovechar del mar, et de su ribera pescando, et navigando el faciendo hi todas las cosas que entendiere que a su pro serán. Empero si en la ribera de la mar fallare casa ó otro edificio qualquier que sea de alguno, no debe derribar nin usar dél en ninguna manera sin otorgamiento del que lo fio ó cuyo fuere, como quie que si lo derribase la mar, ó otri, ó se cayese él, que podrie quien quier facer de nuevo otro edificio en aquel mesmo lugar.
12. *Ley VI Como de los rios, et de los puertos, et de los caminos et de las riberas pueden usar todos los homes comunalmente.* Los rios, et los puertos et los caminos publicos pertenescen a todos los homes comunalment, en tal manera que tambien pueden usar dellos los que son de otra tierra extraña como los que moran et viven en aquella tierra do son. Et como quier que las riberas de los rios sean quanto al señiorio de aquellos cuyas son las heredades á que estan ayuntadas, con todo eso todo home puede usar dellas ligando á los árboles que hi estan su navíos, et adobando sus velas en ellos, et poniendo hi sus mercaduras; et pueden los pescadores poner hi sus pescados et venderlos, et enxugar hi sus redes, et usar en las riberas de todas las otras cosas semejantes destas que pertenescen al arte ó al meester por que viven.
13. *Ley IX De quáles cosas pertenesce el señorio et el uso dellas comunalmente á todos los homes de alguna cibdat ó villa.* Apartadamente son del comun de cada cibdat ó villa las fuentes el las plazas do facen las ferias et los mercados, et los logares do se ayuntan á concejo, et los arenales que son en las riberas de los rios, et los otros exidos, et las correderas do corren los caballos, et los montes et las dehesas et todos los otros logares semejantes destos que son establescidos et otorgados para pro comunal de cada una cibdat, ó villa, ó castiello ó otro logar; ca todo home que fuere hi morador puede usar de todas estas cosas sobredichas, et son comunal

á todos, tambien á los pobres como á los ricos. Mas los que fuesen moradores en otro logar non podrien usar dellas contra voluntat et defenimiento de los que morasen hi.

14. *Ley X Quáles cosas pertenescen á alguna cibdat, ó villa ó comun, et non puede cada uno dellos apartadament usar de ninguna dellas.* Campos, et viñas, et huertas, et olivares, et otras heredades, et ganados, et siervos el otras cosas semejantes que dan fruto de sí ó renda, pueden haber cibdades et las villas, et como quier que sean comunales á todos los moradores de la cibdat ó de la villa cuyas fueren, con todo eso non puede cada uno por sí apartadamente usar de tales cosas como estas. Mas los frutos et las rendas que salieren dellas deben seer metidas en pro comunal de toda la cibdat ó villa cuyas fueren las cosas calzadas, ó en tenencia de los castiellos, ó en pagar los aportellados, ó en lasa otras cosas semejantes destas que pertenescen al pro comunal de toda la cibdat ó villa.
15. Page, "Los Ejidos como Espacio Comunal de la Ciuda de Córdoba del Tucumán."
16. **129.** Señalese a la poblaçión exido en tan competente cantidad que aunque la poblaçión vaya en mucho crecimiento, siempre quede bastante espacio adonde la gente se pueda salir a recrear y salir los ganados sin que hagan daño.
17. **130.** Confinando con los exidos se señalen dehessas para los bueyes de lavor y para los cavallos y para los ganados de la carnicería y para el número ordinario de ganados que los pobladores por ordenança han de tener, y en alguna buena quantidad más, para que se acojan para propios del concejo; y lo restante se señale en tierras de labor de que se hagan suertes en la cantidad que se ofreciere, de manera que sean tantas como los solares que puede haver en la poblaçión; y si huviere tierras de regadío, se haga dellas suertes y se repartan en la misma proporçión a los primeros pobladores por sus suertes, y los demás queden para nos, para que hagamos merced a los quedespués fueren a poblar.
18. **131.** En las tierras de lavor repartidas, luego ynmediatamente siembren los pobladores todas las semillas que llevaren y pudieren haver, para lo qual conviene que vayan muy probeídos; y en la dehesa señaladamente todo el ganado que llebaren y pudieren juntar, para que luego se comiençe a criar y multiplicar.
19. **132.** Haviendo sembrado los pobladores y acomodado el ganado en tanta cantidad y con tan buena diligencia de que esperen aver abundançia de comida, comiençen con mucho cuidado y valor a fundar sus cassas y edificarlas de buenos çimientos y paredes, para lo qual vayan apercevidos de tapyales o tablas para los hazer y todas las otras herramientas, para edificar con brevedad y a poca costa.
20. *Titulo XVII, Ley Vii. Que los montes y pastos de las tierras de Señorio sean tambien comunes.* Los montes, pastos, y aguas de los lugares, y montes contenidos en las mercedes, que estuvieren hechas, ó hicieremos de Señoríos en las Indias, deben ser comunes á los Españoles, é Indios. I Assi mandamos á los Virreyes, y Audiencias, que lo hagan guardar, y cumplir.
21. *Ley IX. Que en quanto á los montes y pastos las Audiencias executen lo conveniente al gobierno.* Los Virreyes y Audiencias vean lo que fuere de buena governacion en quanto á los pastos, aguas, y casas públicas, y provean lo que fuere conveniente

á la poblacion, y perpetuídad de la tierra, y enviennos relacion de lo proveído, executandolo entretando que les constare de lo que huvieremos determinado. Y ordenamos, que entre partes hagan en esta materia justicia á quien la pidiere.

22. *Titulo XVII, Ley V. Que los pastos, montes, aguas y terminos sean comunes, y lo que se ha de guardar en la Isla Española.* Nos hemos ordenado, que los pastos, montes, y aguas sean comunes. . . . á todos los vecinos de ellas, que ahora son, y despues fueren, para que los puedan gozar libremente . . .
23. *Ley VI. Que las tierras sembradas, alzado el pan, sirvan de pasto comun.* Las tierras, y heredades de que Nos hizieremos merced, ó venta en las Indias, alzados los frutos, se se sembraren, queden para pasto comun, excepto las deshessas boyales, y Concegiles.
24. *Ley Viii. Que los montes de fruta sean comunes.* Nuestra voluntad es de hacer, é por la presente hacemos los montes de fruta sylvestre comunes, y que cada uno la pueda coger, y llevar las plantas para poner es sus heredades y estancias, y aprovecharse de ellos como de cosa comun.
25. Ricardo Legorreta, "A Modern Space in Mexico City's Historic Center," *New York Times*, December 21, 2005.
26. Akihiro Kashima, Takashi Tanimura, Tsuyoshi Kigawa, and Masao Furuyama, "Influence of the Italian Renasissance on the Town Planning Concept in the Spanish Colonial Laws," *Urban Transformation: Controversies, Contrasts, and Challenges*, 14th IPHS Conference, Istanbul, Turkey, July, 12–15, 2010. Available at www.iphs2010.com/abs/ID372.pdf.
27. *Ley Xj. Que las tierras se rieguen conforme esta ley.* Ordenamos, que la misma orden que los Indios tuvieron en la division y repartimiento de aguas, se guarde y practique en los Españoles en quien estuvieren repartidas y señaladas las tierra, y para esto intervengan los mismos naturales, que antes lo tenían á su cargo, con cuyo parecer sean regadas, y se dé á cada uno el agua, que debe tener, succesivamente de uno en otro, pena de que al que quisiere preferir, y la tomare, y ocupare por su propria auroridad, le sea quitada, hasta que todos los inferiores á él rieguen las tierras, que tuvieren señaladas.
28. *Libro 3, Titulo II, Ley Lxiij. Que dá la forma de nombrar Jueces de aguas, y execucion de sus sentencias.* Ordenamos, que los Acuerdos de las Audiencias nombren Jueces, si no estubiere en costumbre, que nombre el Virrey, ó Presidente, Ciudad y Cabildo, que repartan las aguas á los Indios, para que rieguen sus chacras, huertas y sementeras, y abreben los ganados, los quales sean tales, que no les hagan agravio, y repartan las que huvieren menester; y hecho el repartimiento, dén quenta al Virrey, ó Presidente,la forma en que han procedido. Y mandamos, que estos Jueces no vayan á costa de los Indios, y en las causas de que conociernen, si se apelare de sus sentencias, se execute lo que la Audencia determinare, sin embargo de suplicacion, por la brevedad que requieren estas causas; y si executado suplicaren las partes los admita la Audiencia en grado de revista, y determine lo que fuere justicia.
29. *Ley XV. Que haviendo sembrado, los pobladores, comiencen á edificar.* Lvego que hecha la sementera, y acomodado el ganado en tanta cantidad, y buena prevencion, q con la gracia de Dios nuestro Señor puedan esperar abundancia de bastimiento,

comiencen con mucho cuidado y diliegencia a fundar y edificar sus casas de buenos cimientos y paredes, y vayan apercevidos de tapiales, tablas, y todas las otras herramientas, e instrumentos, que convienen para edificar con brevedad y á poca costa.

30. *Libro 6, Titulo VII, Ley XXVI. Que los pobladores siembren luego, y echen sus ganados en las dehessas, donde no hagan daño a los Indios.* Lvego, Y sin dilacion, que la tierras de labor sean repartidas, siembren los pobladores todas las semillas, que llevaren, y pudieren llevar, de que conviene, que vayan muy proveidos: y para mayor facilidad el Governador dipute vna persona, que se ocupe en sembrar, y cultivar la tierra de pan, y legumbres, de aue luego se puedan socorrer: y en la dehessa echen todo el ganado, que llevaren, y pudieren juntar, con sus marcas y señales, para que luego comience a criar y multiplicarr, en partes dode esté seguro, y no haga daño en las heredades, sementeras, ni otras cosas de los Indios.
31. *Ley XVII. Que las casas se dispongan conforme a esta ley.* Los pobladores dispongan, que los solares, edificios y casas sean de una forma, por el ornato de la poblacion, y puedan gozar de los vientos Norte y Mediodia, vniendolos, para que sirvan de defensa y fuerza contra los que la quisieren estorvar, ó infestar, y procuren, que en todas las casas puedan tener sus cavallos y bestias de servicio, con patios y corrales, y la mayor anchura, que fuere posible, con gozarán que de salud y limpieza.
32. W. W. Hallo, ed., *The Context of Scripture: Monumental Inscriptions from the Biblical World*, vol. 1 (Boston: Brill, 2003).

Sid Goodloe, photo by Cheryl Goodloe

Sid Goodloe

Sid Goodloe hails from Texas and moved into southern New Mexico in the 1950s. He bought a cow-burnt ranch in 1956 and assiduously restored his land to a state of health similar to that which prevailed in the mid-nineteenth century before the southwestern landscape was overwhelmed by homesteading pioneers accustomed to the less harsh environments of the verdant eastern half of America. He has indeed been in the vanguard of ranchers who recognize that stewardship of the land is not only vital to their success as ranchers, but also to the invigorating and sustaining of a system of land ethics that Americans everywhere would be wise to adopt. He was instrumental in introducing Allan Savory to fellow ranchers and in employing Savory's system of holistic range management techniques that have resulted in restoring much of the ranchland habitat of the Southwest to a state of good health. He lives with his wife, Cheryl, in their handcrafted log home on the Carrizo Valley ranch in the greater Pecos River Watershed. At eighty years of age, he still rides horseback through his ranch tending the landscape with love, intelligence, and hard work while listening to the sound of running water along the creek that nurtures a riparian habitat that he gradually restored from the dry arroyo that prevailed over half a century ago. Here, Sid Goodloe reveals the evolution of his point of view over the last half century as an influential innovator within the southwestern ranching community of practice.

Ranching and the Practice of Watershed Conservation

(THE CONTENT FROM THIS CHAPTER IS EXCERPTED FROM AN INTERVIEW conducted by Jack Loeffler on January 15, 2010.)

JL: What I'd like to ask you to talk about is all the aspects of restoration that you've been involved with since you've had the ranch, because you've been doing watershed thinking all your life.

SG: Well, I'll go back to when I got the place in 1956. It was completely worn out, totally abused. It had sixty head of cattle on it, and ten or fifteen of them were yearlings. It was drastically overstocked because there wasn't any grass. It was all juniper and piñon, and the grass was just eaten off to the ground. I bought the ranch in the spring, and it was the end of the fifties drought. It didn't rain that summer until we got three-quarters of an inch about the middle of September. That's the only rain we got, and it greened up a little after that. It was just sittin' there pantin', waitin', you know, and it greened up.

I got through the winter by burning cactus all winter. There was no grass to burn, so you didn't have to worry about a fire. You just burned the cactus. The place had cholla all over it, lots of cholla. I guess it was in the fall that a government program was introduced that was drought relief where we could buy rolled milo for a dollar a hundred. Now that's twenty dollars a ton. We could mix it with salt and cottonseed meal. I can't remember what the price of meal was, but the salt was the controlling agent. I rustled up enough money somewhere to build some troughs. I've still got one of them, by the way. I built a little shed and fed those cattle meal and salt and cactus all winter. That's how I got through the winter.

In '57 it rained and made a little grass. As soon as I got here, even a Texas Aggie could figure out that there was something really wrong. I was fortunate enough to look at some petroglyphs near here, and I saw fish and beaver on the petroglyphs, and I looked at this canyon and it was nothing but boulders. I never heard the word *riparian*, but I took a picture because it had a little water, the first time I'd seen any. In '56, it was totally dry. But the winter of '57 we had enough snow that when it melted it ran in the canyon for a while. The picture shows what it looked like. It was just boulders and juniper along the sides. Now it's got willows and reeds and sedges. I built up probably anywhere from two to four feet of soil because the fact is that the forest was just as bad as my ranch—overgrazed to the point of terrible erosion. I've been getting [a] tremendous amount of free silt to build up my water table because I don't graze my riparian zone in the growing season, and it collects silt. So I've got a fairly decent riparian area, about two and a half miles, where long ago there used to be fish and beaver.

I had no kinfolks here in New Mexico. My dad died when I was sixteen, and I had no kinfolks out here at all, and not being from this country, I had no advice. So I made up my own plan, and I was able to figure out that these watersheds were in terrible shape because all the water that fell went down the creek. You could almost walk on the creek. You could smell it for a quarter of a mile. It was just mud. Every time it'd come a shower up on the forest, it'd run that mud down here. So I decided to go to the Soil Conservation Service to see what I could work out with them to do something. The in thing at that time was chaining. It was cool to chain. And so I found somebody that was doing that, and the fellow that did most of it in this area was named Charlie Crowder. He was quite famous at that time. In fact, he's the first guy that ever brought a D-9 into this country. That was a tremendous monster of a machine.

We got an anchor chain and two Cats, and we chained 1,700 acres. And immediately the canyon started running again and the springs came back. We did it wrong, but we did something anyway. We chained one way, and the Soil Conservation Service cost-shared on that. I drove one of the Cats, so I was able to do it. But the problem was, we quit after we chained it one time, and a lot of these trees were young enough that they just pulled over; they didn't pull up. We also didn't discriminate in some cases in the old patriarch junipers, the two-, three-, four-hundred-year-old junipers. If I had really realized what I was doing, I wouldn't have chained them. But we didn't get them all; we just got a few of them. Still, it was a shame because those are climax residents. They were here when fire was the regime, and it was keeping the population of all the other trees down.

Anyway, we had water then in the canyon. We should have burned after the chaining also, but we didn't do that because fire was definitely a no-no in those days. You just didn't burn. So we learned our lesson. I had to come back in later with a Cat and pile that chained-over brush and burn it because it came back. But it did a lot of good. I imagine we probably got 40 percent or 50 percent of the trees, but the remaining 50 percent came back, and I had to redo it. We learned that the watersheds are in bad shape, so let's do something about it and we'll get our water back. Of course a tremendous amount of wildlife and livestock feed came after that because those soils had been dormant for a hundred years.

A woman from the University of Arizona named Sutherland did a paper on the pioneers' journals, and in those journals they mention the fact that you could drive a wagon with the bows up and the sheet on the bows through this whole country and never hit a limb. And you could see for a long ways because [in] the words they used, "it was unencumbered by brush under the trees." When I bought this place in '56, you couldn't see fifty yards here because of all these young trees. When she wrote that, she got the dates, and that was in the late 1800s when the first pioneers came through and wrote up their journals. And then she went into the weather, and she found out that there was a period between 1916 and 1920 that it rained heavily every year, providing perfect conditions for tree seedling establishment. There was no more fire because the massive movement of cattle came in here about the turn of the century or before.

In fact, Nathan Sayre did a paper on the cattle boom in Arizona. In 1870 there were forty thousand head of cattle in southern Arizona, and in 1891 there were a million four hundred thousand. Then the drought came in 1892 and '93. Three-quarters of them died, but they'd already done the damage. The land was raped by too many sheep and too many cattle. So the rainfall records show that this wet period was 1916 to 1920, and all these seeds that had been layin' there all came up, especially in the ponderosa. There's 420 or 430 logs in our house, and I counted the rings in a lot of them. When you count the rings, they go back to that period.

She had it all figured out. I really need to find Sutherland's paper because that tells the story of this tragedy of our watersheds completely going down the tube because of a lack of fire and prevailing climatic conditions. There was a very wet period that got them all started, and then we had the drought of the thirties, and that further damaged the grass, but the trees held on. Now we have no fire. Controlled, prescribed burning is just an antonym; it's not a fact anymore because the fuel has built up for 150 years. Now it's dangerous. It's dangerous to do prescribed burning. These people have built these houses all out in the trees, and they're just kindling for the fire. Something's got to happen, and it is happening. The 2010 farm bill really emphasized conservation. There's money available now for watershed rehabilitation.

In my opinion, the public owes the ranchers something for taking care of the land, and now with knowledge beginning to infiltrate the younger ranchers, they realize that they were ranching wrong because they were doing it like the old-timers did it, which is continuous grazing with too many livestock. Now we're into rotational grazing and proper numbers, and the ranges are coming back. But we've got to do something about these watersheds, and that's where the public is going to have to share some of that expense.

That's pretty much the story of the watersheds. Basically it was a lack of knowledge by the original settlers that came from a high-rainfall, high-humidity, low-wind area, which is friendly, into a brittle environment, which is low rainfall, high wind, cold temperatures, and low humidity. These early settlers didn't understand. They were too interested in saving their scalp, protecting their families, building a house, building schools, fencing their property. Those things took precedence over everything else, and the grass was just there. And they abused it. You couldn't move cattle in those days with a truck or trailer. If you got stuck on your property with too many cattle in a drought, what are you going to do? You can't sell them. Nobody wants them. There's not enough trucks to move them to Kansas or Missouri. So you're just stuck with them, and they just ruin the country.

One of my big problems has been the fact that the old-timers that were here when I came here are all gone now. They wouldn't admit that their forefathers had caused the problem. Now they're beginning to realize that they didn't do it on purpose. They just didn't know, and circumstances were against any kind of range management. And now thanks to Allan Savory we have a method to rotate these cattle just like the buffalo rotated down through the plains and created the tall-grass prairies and all those wonderful soils because they rotated themselves because of the seasonal differences in growth on the plains. And we can mimic that by rotating our cattle.

I was just lucky. I ran into Allan Savory at Aubrey Mountain in Rhodesia back in the 1960s. When I moved to Kenya I went back down there to check on it because I thought there was something really good about his rotational grazing method. I didn't know enough and was just too young to realize what Allan Savory had come up with. I went back to Rhodesia twice while I was living in Kenya and talked to the ranchers that were using short-duration grazing. Allan had his own plane. He flew me all around Rhodesia, and we looked at all these different ranches and how they'd been—like if they had a five-wire fence they'd take two wires off of the boundary fence to fence their internal pastures. They didn't have the money to buy the wire. Then they began short-duration grazing. Basically it's a lot of cattle for a short period of time on an area with a long period of rest, long enough for the grasses to regenerate.

I decided he had something, so I got all the information I could and went back to Kenya and wrote it up, sent it in to the *Journal of Range Management* in November of 1969. That's when I went back to Texas A&M, when I got back from Kenya, and I started getting letters from all over the world. People realized that this was a sensible, simple way to manage ranges. So now it's evolved into holistic management, planned grazing, all these different things that Allan Savory brought to this country. He did more for range management than all the range managers, professors, government employees put together in all the years behind him.

JL: One of the things that comes to mind in all of this is that the bison rotated themselves. Do they have basically the same effect on the land itself as cattle do?

SG: Sure, exactly the same. The only difference in the bison is that they evolved into an animal that can stand blizzards, the way the hair grows on them and all that. They're massive in front with not much meat in the hindquarters. The cattle we bred up of course are different, but they can't stand the weather like the bison. The bison, through their history on the Great Plains when the big blizzards would come and cover up everything, they sweep with their nose. They sweep with their nose and expose the grass. Have you ever watched one sweep the grass free of snow and then eat it? That's just a natural happening after many thousands of years on the Great Plains. They moved south in the winter as the cold came and then moved north as it greened up.

The thing that's missing today, Jack, is the predator. Predators, these animals were excited a lot, and they churned the soil with their hooves; they cultivated the soil with their hooves. That's what we call hoof action. They urined on it and they manured on it, and they fertilized it just like a crop. That's how the grass became so good on the Great Plains. Of course now it's all plowed up.

JL: Speaking of predators, how do most ranchers feel about having predators these days in this part of the world?

SG: I know how I feel about them. We'll never be able to get back to the actual predator action. In a massive herd of buffalo, the wolves were all around them, and they kept them stirred up, and they'd panic them at night. That's what stirred the soil. That's what ground in the dead grass and put humus in the soil. That's what got the manure and urine in the soil. Then they left it alone. Then it rained on it and the grass grew. But the attitude toward predators today, it's really something that's kind of like oil and water. I don't think we're going to be able to mix wolves and livestock. It's really kind of pathetic what's happening out in western New Mexico. The Defenders of Wildlife claim that they're compensating ranchers for dead animals. Well, number one, it's very difficult to prove and to find the animal. Number two, there's just no way to measure the harassment of these animals and to measure the amount of sleep that's lost by a rancher who knows there's a wolf in their cattle. They don't pay for all those things. They just pay for the dead animal, and that's not fair.

It might work if they could keep the wolves in an area where they wouldn't be harassing the livestock and the livelihood of people who have settled that country and kept it in the United States basically; kept it out of Mexico. If it wasn't for those pioneers and those old hard-headed ranchers, we wouldn't have Arizona and southern California, New Mexico, west Texas. We'd be in Mexico. And nobody gives them credit for that. Nobody gives them credit for their self-reliance and their stubbornness. They all hold it against them, and that's not fair. You start introducing wolves in an area where people for generations have held on to that land, and you create a major conflict.

I have a problem with coyotes. My attitude about coyotes is I do not mind having coyotes here except when they get numerous enough to run in packs and they start killing calves and they kill fawns. They just wipe out our fawns in this country. Fee hunting is a large part of our income here. And they keep the deer numbers low enough that the state game department realizes the deer population has crashed, and they won't admit that number one, it's coyotes getting the fawns and lions getting the bucks. These old bucks, they're solitary. When they get up [to] three, four years old, they're solitary, and that's when they become a trophy animal that's worth a lot of money to a fella who has them. But the lions kill them. They lay down alone, and the lions can find them and kill them. Whereas a young buck is running with does all the time. He's got does that are looking around all the time, and they don't get those. They get the big trophy bucks.

So I've testified, and I saw on the Internet the other day where I testified to the game commission about these lions. They've got a quota. The quota for the southeast quarter of New Mexico was thirty-four last year, and only nine of them could be females. Then when the quota's filled, that's it. No more hunting. Well, the numbers just keep building and building until the deer population's what suffers. They don't hurt my cattle at all. I don't have a problem with them with cattle, with lion or bear, either one. But I do with coyotes if they get in a pack. Once they start packing up,

I'll know it, and I'll get a trapper in here, or I'll do some trapping of my own and I'll get the numbers back down.

One of the things I don't allow here is rabbit shooting. I don't allow my grandkids to shoot rabbits, because they keep the coyotes happy, and the coyotes kill a lot of rodents and stuff, so that's fine. But when the population of coyotes and bobcats builds up, they start working on the turkeys. Turkeys are very important to me. I don't allow turkey hunting because the predator problem with turkeys is really bad, and the turkeys keep this ranch free of grasshoppers. So that's something I try to control if I can, but it's pretty hard to do. It just takes one bobcat to wipe out a bunch of little chicks, and of course they protect the goshawk, and he's a chick getter too. So there's a lot of problems with predators, but you've just got to kind of use a little common sense about populations. Keep them in check, but don't try to wipe them all out.

Now, wolves are a little different. Wolves, it just takes one pack of wolves to upset a whole bunch of ranchers and kill a lot of cattle. There's documented evidence now of how many cattle they kill, so it's a problem.

[Time out for a fine lunch prepared by Sid's gracious wife, Cheryl]

SG: I've been trying to work with the Forest Service and have, in a way in the past, to get them to thin [the trees]—the big part of the watershed here that supplies my riparian zone is on the [National] Forest. It's all above me, and it's overpopulated with trees, so very little of the water gets in the aquifer. So my watershed is a victim of a large area of overpopulated trees, whereas the small area that I've done is a benefit and has helped a lot. But it could be so much better. We could have beaver and fish if they would do to their watershed what I've done to mine and thin it out and make it a productive, functioning watershed like mine is. I've tried my best to get them to do it. They actually have promised me some thinning for the last three years, and they swear they're going to do it this year. We'll see.

JL: This is fascinating. I did a radio program on Aldo Leopold earlier this year. Up where I used to be a fire lookout in the Jicarilla Ranger District was where Leopold was at one point superintendent of the

Carson. And he was riding on a horse from Durango back to Tres Piedras, and somewhere he got stuck in a big storm in the ranger district where I was a fire lookout and got really sick, almost died. It was in that ranger district where he changed his tune about—

SG: Trees and the watershed.

JL: And the watershed, yeah, and overgrazing, because that country had been pretty heavily overgrazed.

SG: Well, we've got all his books here and all that, and if I'd have read his books before I came here I'd have saved myself a whole lot of work and time because I had to figure it all out. I didn't know he had it all figured out. What I learned, he knew back in the twenties. I really admire Aldo Leopold's method of figuring all this out. He really had it nailed.

JL: One of the things that Leopold addresses in *A Sand County Almanac* is how conscience is a factor that's too often excluded when looking at land use.

SG: Yeah, sure!

JL: To me conscience is something that gets overlooked left and right. Obviously you come from conscience.

SG: Yeah, and common sense. It doesn't take a rocket scientist to figure out there's too many trees on this place. I've said to more than one person, I'm really not very creative. I'm too practical. Practicality kills your creativity. They say no, you're creative. You are a landscape artist. I guess that's what I am, because we're clearing a little valley over here now that once we get it back to climax condition in a couple of years, that's going to be the prettiest thing you ever looked at. And that's what we're doing.

My wife Cheryl and I cut about twenty-five cords of firewood and split it and sell it in Ruidoso every year. So I'll cut probably five, six acres is all. But it's so thick; it's just a solid mass of stumps. You can't hardly drive in there to get the wood, they're so thick. I leave all the old matriarch, patriarch—whatever you call them—junipers. I leave them. That's where the good wood is. But we cut up the smaller trees. Gee whiz, it's a great reward to me to go into a place that I've restored and reseeded that's come back and looks like it did 150 years ago. That's quite a treat for me.

JL: What species of grass?

SG: We try to use native species, but there's a couple of things that I do that are probably not very ethical. One of them is that I learned over the years that this Paiute orchardgrass, which is not a native, was developed in Utah to stand this sort of climate. It's a real attractant to elk, and I get good money for my elk permits. Before I got permits, I quit putting it in the mix because it attracted the elk, and they just hammer the heck out of this ranch. But now that I get the permits, I want the elk. I've got too many, but we're getting them thinned out a little bit, maybe. Another grass that I've been using is weeping lovegrass because it is so hardy, and it will come immediately and stop erosion. If you've used a Cat on a hillside to clean up some old multistem junipers and all, the first year it's susceptible to erosion. But if you use this weeping lovegrass, boy, it'll come in there, and the cattle and the game love it. A lot of people don't like it, but I do.

In most areas we've got plenty of blue grama seed anyway. I use western wheat, sideoats grama, galleta, and I'll put in some clover. That helps a lot. That attracts the deer and helps hold the soil. Basically the western wheat, blue grama, sideoats grama, galleta, those kinds of grasses are what I use.

A lot of people think the elk have run the deer off, but I told you what I think. It's different. I'll tell you something else that's happened. The game department's finally realized the importance of habitat, getting rid of all these thickets of juniper, then the cows can't get the fawns, and the lions can't get the bucks if it's opened up. That's why I've got deer here when they don't have them in a lot of other places where they haven't thinned the trees. So it's finally dawned on the game department that habitat is a big factor. And because I've done all this habitat work, I get two extra trophy hunts. There's I think thirteen ranches in the state that get them because of habitat work, and we're one of them.

JL: That's amazing. You mentioned fire earlier on. I know that down in the Malpai Borderlands Group—are they still using fire?

SG: Oh yeah. We use it here. It's just that, here's the deal. The county commission is responsible, so they put a fire ban on in the spring as soon as the snow goes and it gets dry. There's a fire ban put on, and it doesn't go off until it rains for two weeks. And so the period that Nature always took care of these trees is in the spring, and you can't burn. Plus the fact that if you are gonna burn, even if they didn't put a ban on it and you could burn, what rancher's got enough equipment and personnel to do a fire? So what I've tried to do is get the state forestry, the U.S. Forest Service, and the local fire departments to all get together for training. I've offered the ranch. Get together and do some training on putting out these grass fires, and let's burn. And I can't get it done. So we just have to be really careful, and what we do is basically our burning now is limited to burning slash from the fuel wood cutting.

As I said previously, the big problem when people came to this country was survival and total ignorance of a brittle environment. Those are the two things that caused the demise of our watersheds. A brittle environment that was low rainfall, low humidity, high wind, cold in the winter, hot in the summer—and those early settlers just didn't know how to handle the climate. They thought they could handle it the same way they did back in Louisiana and east Texas and Oklahoma. Consequently, the degradation was contributed to not only by their lack of knowledge but by climatic factors, because we had some terrible droughts. That's one of the things they didn't understand. We are going to have droughts in this country, and we'll go for maybe two or three years without any significant rainfall. In your planning of grazing the range lands here, you've got to have a drought valve somewhere in your operation to fall back on, and contrary to what their conditions were where they couldn't move cattle by truck and get them off of a drought-stricken ranch, we can do that now.

But we've got to plan for that. We've got to have that as Plan B, to be able to do that. So as time has gone by, the ranchers, maybe three generations of ranchers, have finally realized that they can't graze this country the way their forefathers did back in Louisiana.

And so the management has improved and the consciousness of a sustainable operation is now foremost in their minds. They realize that they can't damage the entity that they make their living on. Their income drops as the condition of their land drops. They've realized that for the most part—not all of them, but a majority have done that.

Plus, we have another factor that has entered into the overall scheme of things, and that is that in many cases the children and grandchildren aren't interested in staying on the property. Consequently, when maybe the original owner or one or two generations down from the people that settled this county, when they die, two things happen. One is that the kids aren't interested anyway, and the other is the fact that death taxes require a 55 percent tax on the estate. The kids cannot pay that without selling the property, and the buyer that shows up is a developer, because he can pay more money than anybody else.

If somebody can step in between the children and their tax problem and the developer and buy that ranch and then use common sense and education to manage that ranch in a more modern approach to grazing, which is short-duration grazing or planned grazing, then things begin to improve rapidly. They have a more environmentalistic approach to management than the old-timers. The old-timers didn't have time for the environmental thing, and then about 1970 when Earth Day originated, here come all these graduates out here telling all these old-timers that had roughed out a life out here, and break even only without any money, who'd just barely hung on, they come out and tell them to change their operation? Telling them how to run their business? A twenty-year-old telling a seventy-year-old rancher that's been here all his life how to run his ranch? That did not work, and that's what built the wall between the ranchers and the environmental community.

Now with these new people that have come in that have a more environmental outlook and younger people that have gone to school and been exposed to environmental sensitivity, then things are beginning to improve, and they have a conscience that prohibits them from abusing the land. They get great gratification out of improving the land and bringing it back, at least in my case, to a presettlement condition. I may never get there, but at least I'm headed in that direction. At least I've noticed a lot of other ranchers are headed in that direction. They may not admit it, but they are, and it's just a survival, sustainable approach.

JL: Right on the mark. I would like to contribute to bridging the gap between the ranching community and the environmentalist community.

SG: That's what Quivira Coalition was formed for. What they're good at is bridging the rural-urban divide and getting ranchers to understand the environmental approach to ranching. It's not a strict environmental tree-hugging approach. It's just a little sensitivity to getting a little biodiversity in your rangelands, to bring back your watershed, to run the cattle that fit the environment. Those kinds of simple things are now beginning to be received by mainly new ranchers. In fact, the Quivira Coalition has a term, *new ranch*. That's the new approach to ranching. Well, that makes a lot of people kind of look at it crossways, but it is. It's a new approach. It's an environmentally sensitive, economically solid approach to ranching.

JL: Sid, talk about SUVs.

SG: Well, ORVs is the way this all started with me, off-road vehicles. That includes pickups. This was before the SUVs came out, before the four-wheelers. This was just four-wheel-drive pickups. We had a tremendous problem with them here because there's Forest Service all around me, and they made roads, and the roads eroded, and it was like the old roads from the homestead days. They were just gullies. They would road-hunt out of them all the time. So I was on the user's advisory board to the secretary of agriculture for six years, and we met in Washington every year as one of the meeting places, and every time I'd go to Washington I'd stay a couple of extra days and make appointments with like the Audubon Society and Natural Resources Defense Council, all those environmental organizations. I'd go around, and I had a slide presentation on off-road vehicles that you wouldn't believe. Nobody could deny the damage that they did.

These people were not at all interested then. Finally the Users Advisory Board did a white paper for the secretary that expounded on all the problems with these deals. It took twenty years to get any interest in this, and now it's beginning to show up in the Forest Service—they've got a new project that they've had on designating roads and things like that to try to stop off-road vehicles. Simply because I started this back in the '80s, they kind of got ahead of everybody else—they have declared off-road vehicles illegal.

We have a guy here that enforces it pretty good, but it took twenty years to get that done, and the BLM [Bureau of Land Management] is now trying to get a travel policy, and their policy's always been that the BLM lands were open unless designated closed, whereas the Forest Service was closed unless designated open. But with the lack of enforcement, what good did it do? But the Forest Service has done a lot here. The harassment to the wildlife and the damage to the land are just inexcusable, and the manufacture of four-wheel-drive vehicles just like the one you came in is getting to be common procedure for manufacturing anything that's a pickup or an ATV or SUV or whatever. They're all four-wheel-drives; they can go anywhere they want to go. They love to go out in the mud and cut ruts, and the ruts become gullies, and the population is exploding just like the tree population did a hundred years ago.

Somebody better step up to the plate and put some regulation on these off-road vehicles. Either that or you're going to have to declare all this public land a wilderness area so they can't go in there at all. If they'll just designate the roads and enforce staying on the roads, then okay. But they rarely do.

JL: Now I really want you to address conservation easements if you would, what that really means. Define them.

SG: Well, a conservation easement, number one has got the wrong terminology. To a rancher the word *easement* is a red-flag issue because it's a power line or a pipeline or an oil field road or an access road or whatever, but the word *easement* is starting off [with] two strikes against it. It should be a "conservation agreement." This is a document that the landowner uses as a guide for the things that are absolutely required by the government—the government being the IRS. If you want to use the conservation easements for any tax benefits, then you have to follow IRS regulation just like you do with your own business. So when the landowner talks to his family and decides that they want to do a conservation easement, which in effect lowers the value of the property because it can't be subdivided, that way then you can get around the death tax or the estate tax. That's the main reason.

Now there are more benefits that I'll get to in a minute, but basically you decide on whether or not you want to do it with your family. If you want to leave your family financially well off, you don't have any business talking about a conservation easement. A conservation easement is for people who love the land and want to keep open space, wildlife habitat, and family agriculture going. Those are the three things that we try to do here, plus protecting historical sites. Once the landowner and his family agree that they want to do a conservation easement, then they get a blank document, and in that document it says that there are certain uses that are prohibited and certain uses that are allowed. You put in there what you want prohibited and what you want allowed, so in effect you're ruling from the grave.

There are conservation values if you're going to do a tax benefit easement that you have to observe. You have to protect open space, you have to protect agriculture, have to prevent development, and you have to protect wildlife habitat and historical sites. Any one of those will qualify you for tax benefits under the IRS. I'll get to the state benefits in a minute. Then once you come up with a document that you've talked to your accountant about and you've talked to your lawyer—and you need to be careful about a lawyer because if you use the family lawyer he will tell you not to do it because he knows nothing about it. Conservation easements are too new in this part of the world. They're not new back east, but they're new in southern New Mexico.

So after you talk to your accountant and your lawyer and you get all the legal ramifications and figure out what you're going to do about tax benefits, then you have to get what we call a baseline inventory. We basically did this with Dr. Jack Wright from New Mexico State. You can do it on your own, but it's better to get somebody else to do it. We're talking about money now, and all these things are going to cost you money. So what you do basically is you record all the plant species and animal species that are on the ranch at that time. You have photo points all over the ranch so a hundred years from now you can go back and look at those photo points and see if you've allowed the brush to take over or if you overgraze or erosion's been present or whatever. You have a record of not only the land and the animals but the buildings, all of the buildings, and the roads.

In the first place, you've got to remember you can do any part of the ranch. You don't have to do the whole property. You can reserve a building envelope around all the buildings. This house has a five-acre building envelope around it, so I can put another house on it if I want to. Down below is ten acres so I could put in a tourist lodge or a wildlife photography lodge or whatever. You can do all those things, but they have to be in the document, and the baseline inventory has to have photos of what exists at that time.

Then the last thing you have to do is an appraisal. You get an appraisal of the ranch as an agricultural property as it is, or as what they call a fair-market value, which is with the potential of development. Let's just use a hypothetical case. If this ranch appraises at fair-market value at $2 million and as an agriculture property at a million and a half, then the value of the development rights is that $500,000 between the million and a half and the $2 million. The two different appraisals give you a value of the development rights, and the development rights are retired. When you do an easement, you finish your document, finish your appraisal, finish your baseline inventory, get everything ready, and you go and record it in the county clerk's office. Once it's recorded then it's with the deed from then on. It goes with the deed. You can sell the ranch; you can lease it; you can do whatever you want to do with it, except abuse the conservation values that I mentioned or subdivide. Those are the two limitations. And because those limitations are on there, the value drops. So when you die, unless it's a great big ranch, you probably would fall under some kind of an exemption. If you didn't, then you would have much less tax to pay.

The reason that conservation easements are so popular right now in New Mexico is because the legislature three years ago passed a law that you can have state income tax benefits—if I donate an easement worth $500,000 to a land trust, then I've got a $500,000 income tax benefit. One individual can only have $250,000, that's maximum. A couple can have $500,000. So if we had done this after the law passed we'd have a $500,000 benefit from state income tax.

I don't need a tax benefit. I don't make enough money to pay significant taxes. So what good is the state income tax benefit to me? So the legislature passed a bill that you can sell those benefits. That's called a state income tax transferability clause. I've got this $500,000 credit, and the law that passed says that I can sell that over a period of sixteen years. I've got sixteen years to use it or sell it. So if I say I need the money now, I'll sell it. I go to a corporation that owes a million dollars in state income tax, and I say, "I've got $500,000 of credit here I'll sell you for $450,000." We both made money. That's why conservation easements are and should be even more popular in the state than they are. It's just because of the inherent fear by ranchers that was planted by the Paragon Foundation and others that make them think the government's going to take over their land. And it has nothing to do with the government, except the IRS, and we all deal with that.

Basically what a conservation easement does besides the tax benefits is it tells you what you can do with your property when you die rather than the IRS. Because if you die without any adjustment at all, land values in New Mexico have tripled in the last three years. So you're going to pay 55 percent of the value of your estate, which is the land valued at a development-type appraisal—that could be developed, that's a high appraisal—all your insurance policies, all your equipment, everything you own is totaled up and you pay 55 percent within nine months. That's the law. So a conservation easement to me is a no-brainer. Cheryl and I did it five years ago on this ranch. So if I die this year, no problem. But if I wait till next year, we go back to the 55 percent. Now how in the hell are these kids going to pay that? And why would I be taxed after I die when I paid tax all my life? On everything I buy I pay tax. Everything I sell I pay tax or use it as a write-off for the ranch. Gee whiz, that the most unfair tax in the world is an inheritance or death tax. All it is is South American-style land reform without the guns. That's all it is.

JL: Then the other thing you were going to mention is you were going to talk about the archaeology of this watershed and land.

SG: Most of this I've already said, but in this country, the archaeologists tell me there were 1,400 Indians living down there in that village down there by that house. I don't believe there were that many, but there were a lot of them. There's a lot of houses still down there. They had fish, and they had beaver, and they had game. They had all these things that went with a functioning watershed. That's why they were there, because the watershed functioned and there was water. Water is the key to this whole country. That's the way it was settled, by the first livestock people, sheep, and cattle. If you controlled the water, you controlled all the land around it that you could protect with your six-shooter. So when a watershed is functioning, then the water's going into the aquifer and showing up in the springs and the creeks. The surface water comes from the uplands in the watershed that are now completely devoid of any percolation of water into the aquifer because the solid canopy of trees gets all that water before it gets into the aquifer.

They say if you have a good snow, then that snow will fall and it will go into the ground before the trees start using it. Number one, in this country a large part of the snow that falls on these solid canopies is evaporated before it ever gets to the ground. I've been cutting trees here for fifty years or more, and I guarantee you that if you go and cut a piñon right now, in a week's time there'll be a layer of sap on top of that stump. Those trees are using water in the wintertime. A juniper tree, if you have enough moisture in the ground, and the roots go down 150 feet, and there usually is water somewhere, if you've got any water and you get a warm spell in the winter, it'll put on another ring. That's why you can't tell the age of a juniper tree. Ponderosas don't do that. Piñons don't do that. But juniper do put on another ring. So if you've got a high population of trees, for instance, if you have 100 six-inch juniper trees per acre, and that's not a whole lot, they use nine inches of precipitation a year. A six-inch juniper tree will use an average of seventeen gallons of water a day yearlong. That's a lot of water. So none of this water reaches the aquifer, and the aquifer is the secret to survival in the Southwest. You've got to have surface water if you don't have a drilling rig, and the Native Americans didn't have a drilling rig so they lived on surface water.

JL: Do you know what tribes were down there?

SG: This was a mixing area. They called them Anasazis, which means *ancient ones* and all that. But they were the Mimbres and the Waco and I forget what else. But according to the archaeologic book that was written on this place, there were three stages of Native Americans here: the pit dwellers, the rock dwellers, and the adobe dwellers. They followed each other. You can see the evidence down there.

With the history I've given you about the rainfall and the [period from] 1916 to 1920 and all, it all matches up. See, Jack, everything has a predator, even elephants. Lions will get the little ones. The predator that kept the tree population under control was fire, and it's gone.

When I cut firewood to open up this country—I use three things to bring this watershed back. The bulldozer, when the thickets are so thick, then there's no value in them at all. I doze 'em up in a pile, go right behind the dozer and put out native grass seed, and then a year later I'll burn the pile, and then I'll reseed that spot and I'll get a good stand of grass. The other way is to cut them with a chain saw and use the firewood to pay for it. The third way is with an herbicide. Actually I was putting out herbicide yesterday. The reinfestation is the problem. If you do any of this watershed rehabilitation without a maintenance clause in it, you're wasting your time. It'll come right back. There's so many vectors now: birds, coyotes, turkeys, cattle, horses, everything eats juniper berries, and they scatter them with their own manure. They've got fertilizer and everything. All they need is the right kind of year, and they'll come up and they'll come up everywhere. They continually come up, and there's no fire to kill them after they get up six inches tall. They just keep growing. The agriculture research service in Phoenix has proven that the increased carbon dioxide in the atmosphere is causing these trees to grow faster than they have grown since the age of the dinosaur. We've doubled the carbon dioxide content in the air in the last hundred years, and that has, according to their research, tripled tree growth. It doesn't seem to do much for grasses, but it's tripled tree growth.

I've been here long enough to tell. I can look at these trees and tell: they're growing a lot faster than they used to. They've got a different shape. So what we've got is a watershed that can be rehabilitated, but if you don't maintain it—if you're going to do a government cost-share to help these ranchers do this, to help the watershed, to help the aquifer, to help the cities that use the water out of the aquifer, then you better have a maintenance cost-share written into the deal. Don't do just a cleanup, do a cleanup and a maintenance, or don't do anything.

It's the same thing with the rivers as it is with the mountains. The rivers are totally consumed by salt cedar that's using all that water up, the same as the ponderosa and P-J [piñon-juniper] up here. They've got to get rid of that salt cedar, and also they've got to get rid of a solid canopy of P-J. Now I've got the data that shows that the Pecos River measured at the weir in Artesia has lost over 30 percent of its flow due to vegetative change on the watershed.

One of the things that comes to mind is the concentration of roofs and pavement in Ruidoso. That's got to affect the watershed, because none of that water goes in the ground. If they ever get the right kind of rain like they [did in] July of '08, look what happened. My goodness, there's maybe one-tenth of the amount of land available now over there in that area to absorb water and the rest of it, down the creek she goes. As long as we keep putting subdivisions on the watershed, we're going to change the structure of that watershed just like carbon dioxide is changing the way these plants grow. So to me a pretty important consideration in any watershed is the number of houses and pavement that are in that watershed.

Celestia Loeffler, photo by
Michael French

Celestia Loeffler

CELESTIA LOEFFLER, COPRODUCER OF *WATERSHEDS AS COMMONS* AND LORE of the Land board member, was born and raised in Santa Fe, New Mexico, where she quickly realized the magic of the region. She has since dedicated her life to the art of cultural and environmental preservation and has crafted a unique career as a producer, archivist, model, wordsmith, and yoga instructor. She is currently working with Lore of the Land on myriad projects that promote and preserve the cultural diversity of her beloved homeland. She holds a BA in creative writing from the University of New Mexico and is certified to teach English as a second language through the School for International Training. She makes her home with her husband and the surrounding wildlife in Durango, Colorado.

Roots of Hunger

The Quest for a Sustainable Food Culture

The fate of nations hangs upon their choice of *food.*
—Jean-Anthelme Brillant-Savarin

We depend on food to nourish and sustain us. As humans have evolved, so too have our methods of feeding ourselves. Early hominids scavenged to find nourishment, relying mostly upon the food they could gather and carry with them from one place to the next. But as we became less nomadic, staying in one place for longer periods of time, thus began early forms of agriculture: the scavenger diet was gradually replaced by domestication of livestock, planting seeds, tending the crops, and harvesting the bounty for sustenance. Fast-forward a few thousand years, and the essence of agriculture remains intact, but the methods and fruits of our labor have changed dramatically.

Over the last few generations, something has shifted in the way we humans—particularly in developed countries—eat. We lead busy lives and have jobs, families, and hobbies that require our attention, time, and finances. Rare is the opportunity to cook, let alone enjoy, a meal that has been thoughtfully chosen and prepared. Not many of us have the luxury to grow our own food, or even know the circumstances under which it arrived on our plate. Was this grown sustainably? Harvested humanely? How much energy did it take to get from its point of origin to my kitchen?

But we all eat. And by doing so without reflecting upon these questions, we are bereft of a sustainable food culture. The result is the ailing health of our planet. Earth is being polluted, its resources exhausted without ample time to repair and replenish. Overfarming and pesticide use are destroying the soil and water systems we rely on for our survival; biological diversity is threatened by genetically modified seeds and monocropping farmlands; the oceans are overharvested to feed the human hankering for seafood; and the majority of animals bred to nourish us suffer horrifying living and dying conditions brought on by factory farming. Our blind, insatiable hunger is killing the planet. But by exploring our channels of nourishment, we might just be able to identify, and even heal, the wounds that reach so deeply into our food systems.

A *foodshed* is a similar concept to that of a watershed. If a watershed follows the flow of water supplying an area from headwaters to spout, so a foodshed outlines the flow of food supplying an area from farm to fork. Having once relied on local land and water supplies for sustenance, we now rely on food (and water) supplies from hundreds, and often thousands, of miles away.

Ethnobotanist and author Dr. Gary Paul Nabhan is a longtime farm and food activist. In an interview conducted by Jack Loeffler in the summer of 2010, he had the opportunity to explore the root causes for our alimentary oversights. He contends that we humans have become utterly disconnected from our food supply, the ramifications of which have had a deep impact on our environment and identity.

"We are connected to a watershed or foodshed just like we are connected to our mother by an umbilical cord," says Dr. Nabhan. "It's what nourishes us. And we have been severed from that nourishing element by the globalization of food systems." Globalization has erased the boundaries of foodsheds previously dictated by Earth's natural food-growing capabilities and limited transportation options. But now, due to the furtherance of farming and transport technology, we in the Southwest can eat fresh strawberries in the middle of January, with two feet of snow on the ground and not a strawberry plant in sight—a luxury not afforded to most of our ancestors.

Nabhan reminds us that foodsheds used to have multigenerational feedback loops intended to keep humans good stewards, wherein any damage to a foodshed or watershed would be healed and repaired within a generation to ensure that the next could eat. Today, those connections are all but lost. As we deplete the soil and water sources one generation after the next, the productivity and carrying capacity of our foodsheds greatly diminish, leaving us all, present and future, hungry.

We have become so distant from the sources of our sustenance that the average food now travels at least 1,200 miles before reaching our pantries. According to Nabhan, almost half of the human carbon footprint, which is proving to be quite costly to environmental health, is actually a carbon "food print." How food is grown, harvested, processed, packaged, delivered, prepared, and eaten has come to require colossal, unsustainable amounts of fossil fuels, which is contributing to rapid changes in the global climate.

Average temperatures are rising in some places, and cooling in others, while precipitation is becoming increasingly variable. Mountain snowpack is melting earlier and more rapidly each year, while higher overall temperatures are causing water stored in reservoirs to evaporate with startling speed. And along with many other systems of practice, agriculture is suffering the consequences. All the while, Earth's human population steadily grows. At the beginning of the twentieth century, the human population of Earth hovered at just over one and a half billion. Today that number has more than quadrupled. Between climate destabilization and the consumption associated with population growth, this is not a sustainable model for human life on Earth.

But there is hope. By reexamining and reconfiguring our food choices, we still have a chance to significantly lighten the burden we've put on our planet and all of its complex and inextricable systems.

Environmental, social, and economic sustainability are intertwined, and none of these issues can be resolved without addressing the effects of our eating habits. By reconnecting with our food, and foodsheds, together we can begin to heal the environmental, social, and economic conundrum that has arisen due to our insatiable appetites. The following is a chronicle of how I, a humble epicurean, have begun to reconnect with my food.

Soil

It's amazing what you find when you start digging: rich soil, rocks, red-brown earthworms, occasional relics such as a metal stake or ceramic shard from those who dug here before me. Each shovelful unearths a clod of history.

I started digging about a year ago, in a lush field in the Animas Valley in southwestern Colorado, nestled at the base of the San Juan Mountains. The Río de Las Animas Perdidas, or "The River of Lost Souls," runs through the heart of the valley and provides the lifeblood, thousands of acre-feet of pristine water, to hundreds of farmers and ranchers to grow their livestock and crops.

Jutting pine trees, red rocky cliffs, and crisp blue skies witnessed as my shovel first touched the earth one chilly May morning. The ground had begun to thaw, and tendrils of new life had sprouted through, excited to revisit the world above. Brave fruit trees bloomed, and grass turned from strawlike yellow to raw, tender green. An occasional crocus, tulip, or hyacinth poked out from the cold soil, only to be gnawed to the nub by a doe and her fawns looking for sustenance after a long winter in the higher elevations.

The Animas Valley is a picture postcard of simpler times. County Road 250 winds along the east side of the valley, past pastures full of cattle and horses and an occasional herd of elk. Plump men in overalls balance pails of frothy milk, fresh from the udders of their dairy cows, with spritely calves in tow. Chickens waddle from coop to barn, scrounging for scratch, and ornery farm dogs yip and bark at passersby. Vegetable gardens dapple the hillside, between farmhouses and barns and an occasional windmill. People here, at least a few, still grow, harvest, and preserve a portion of their own food. Even some young folks, real estate agents or bankers by day, are secret gardeners by night, managing their crops, coops, and hives after a long day at the office. It must be a sign of the times, but most folks have to keep a day job in order to sustain their food-growing habit.

In his book *Bringing It to the Table*, Wendell Berry defends the family farm, which he defines as "a farm that is small enough to be farmed, and *is* farmed, by a family . . . that does not destroy either farmland or the farm people."[1] By extension, the family farm oughtn't harm, but live in harmony with, the surrounding wildlife. Berry contends that the way we farm affects the local community *and* the local economy and that the local community and economy affect the way we farm. These systems are interdependent, and their health and integrity require a healthy ecosystem. If any of these factors are in ill health, then *all* are out of balance, which explains our current agricultural instability.

Healthy family farms are disappearing at a rapid rate. They are being sold to the highest bidder, subdivided, and turned into townhouses and condominiums, or pillaged for their natural resources. The land and water systems once used to provide sustenance and a healthy agricultural counterpart to the local natural ecosystems are being squeezed out of existence for economic gain. There are many reasons for this disconnect. Wendell Berry might blame corporate economics or the mechanization of the agricultural practice. Others might implicate capitalistic individualism that has come to embody America and its inhabitants. The common thread, however, is greed.

THE SOIL IN THE VALLEY IS ALKALINE AND HARD BUT DENSE WITH NUTRIents. Lime, zinc, and copper contribute to the rich red hue of the earth. Following the advice of Japanese farmer Masanobu Fukuoka, author of

One Straw Revolution, I had made the decision not to rely on tractors or tillers to unlock the benefits of the soil. Working the earth by hand, and perhaps an occasional shovel, rake, or trowel, I hoped to develop a stronger and deeper connection with the land.

To this day when I dig, I ruminate on the poignancy of Wendell Berry's observation that "if we work with machines the world will seem to us to be a machine, but if we work with living creatures the world will appear to us as a living creature."[2] I cannot, personally, see any other way to view the Earth but as a living, breathing, sentient being. By dirtying our fingers with pungent earth every time we pluck an edible jewel of the harvest, we begin to reconnect with the land that sustains us.

Mr. Fukuoka believed that "natural farming proceeds from the health of the individual."[3] That healing of the land and the purification of the human spirit are the same process. When we change how we grow our food, we change the food, and by changing the food, we change our values as a society. And even though I didn't follow all the principles of his farming practice, I relied only on my energy, and the occasional help from a friend or family member, to dig into the soil's riches, plant seeds, and tend to the crops.

I built raised beds out of lumber, nails, and chicken wire and amended the soil with mushroom compost and soil-building amendments to house a variety of vegetables and flowers; dug a mound to plant squash and a straw bale bed to sow potatoes; and dug a small raspberry patch to house some meddlesome raspberry runners that were taking over a neighbor friend's yard. Many afternoons were spent digging, hammering, lifting, and building, and by the end of each day I'd be covered in sweat and soil, grass stains, and splinters, with sore muscles and a happy heart. Armed with little more than a shovel, a few dollars for supplies, and a healthy dose of determination, the farming season had begun.

Seeds

Seed catalogs are among my favorite reading material. Perusing page after shiny page of glowing produce, reading lengthy descriptions of a given species' history, cultural requirements, and projected yield, I've been happily seduced into dog-earring, starring, and circling my way through stacks of garden catalogs. But I recently made an alarming discovery.

As long as people have been planting seeds for sustenance, they've bred and hybridized seeds, isolating and nurturing the biological traits most likely to thrive in a given homeland. But within the last few decades, the process is being sped up and manipulated to dangerous proportions. Biotechnologists have found a way to take genetic material from one organism and inject it into another, resulting in a genetically modified organism, or GMO. Genetically engineered seeds have since overtaken farming culture and industry, providing us with such novel creations as cold-resistant tomatoes, which share genes with flounder—yes, the fish—and potatoes that share genes with bacteria. Foods that have been genetically altered have already seeped into our grocery stores and pantries, unbeknownst to us, the consumers, because there is still, as of this writing, no legislation to label genetically modified foods.

According to the Center for Food Safety, the top ten seed companies now own 57 percent of the global seed market, which is creating a tidal wave of negative effects. Many of the large seed corporations are producing genetically engineered seeds to be used in conjunction with toxic pesticides, both of which are being proven harmful to the environment. Also, by covertly reducing and eliminating "conventional" seeds, the corporations are leaving farmers with little choice but to buy and plant their patented varieties.

Certified organic seed companies, which have signed a petition to not knowingly sell genetically modified seeds, provide one of the only safe ways to buy seeds. And even then, due to the inevitability of crosspollination, fields of certified organic crops are being contaminated by the genetically modified varieties from neighboring farms.

Genetically modified seeds and foods have been touted by many, including scientists, as a powerful and safe way to improve agriculture and feed the growing numbers of densely populated starving nations. However, the integrity of GMOs has come into question. Studies conducted within the last decade have revealed that genetically modified foods and other organisms present great risks to humans, animals domestic and wild, and to the environment. According to the Center for Food Safety, "Human health effects can include higher risks of toxicity, allergenicity, antibiotic resistance, immune-suppression and cancer."[4] The full gamut of environmental consequences of using genetically engineered seeds in agriculture is impossible to predict, but of central concern is the uncontrollable biological contamination of plant and animal species with potentially hazardous genetic material.

An increasing number of factory farmers are planting monocrops of genetically engineered corn, soy, canola, cotton, sugar beets, and other plants, wiping out much of the biodiversity to which our environment is naturally inclined. According to Dr. Nabhan, that winnowing diversity is not going to help us in these times of climate instability.

Nabhan suggests that we need as much diversity in adapting to the changing climate as possible because we don't know what's going to make it through this enormous transition. "One GMO variety of wheat or soybeans or corn cannot adapt to the range of local and regional climatic change that we're already seeing," says Nabhan. "They can put 20 million acres into one variety, but in some of those places it will be too much drought, in other places too much flood. We're much better off empowering the incredible diversity of seed stocks and livestock breeds that our ancestors and our neighbors have engendered over the last ten thousand years, than thinking that one silver bullet will get us through this era of rapid climate change."[5]

Seeds sustain the essence of life. It is deeply disrespectful to meddle with Nature by fiddling with these precious harbingers of life. So after a bit of research, I found a nearby organic seed company from which to buy healthy, nongenetically modified seeds to plant in my freshly dug beds.

Before the package seeds even arrived I set to work drawing up a garden plan: chives, spinach, parsley, sunflowers, and three varieties of lettuce in one bed; sugar snap peas, carrots, arugula, broccoli, and lemon cucumbers in another. Beets, turnips, radishes, cilantro, and rainbow chard in the next. The rest of the space would be devoted to bachelor's buttons, California poppies, and other flowering herbs to attract honeybees and other pollinating wildlife.

I made newspaper seed-starter pots, filled them with sterilized potting soil, and seeded my transplantables, keeping them moist in the windowsill and waiting for the cotyledons to unfold out of the soil. In the garden beds I planted some varieties of lettuce and root vegetables, the crops hearty enough to live through another light freeze or dusting of snow. By the end of June, the crops had been planted.

Worms

"If you have to believe in something, believe in the magic of earthworms," my dear mother said to me one afternoon on the phone. She described wandering outside to her small garden in northern New Mexico, stopping at the compost pile, and discovering that after a few short months of working their magic, the earthworms she'd added to her compost pile had created a thick layer of rich, black soil.

Composting occurs in the natural world, without any assistance from humans. In the fall and winter, plants and trees drop their foliage to the ground, which decomposes and is then eaten and "redistributed" by animals, birds, worms, insects, and microorganisms. This cycle creates dense, nutrient-rich soil and can be applied to garden composting. By saving kitchen and garden waste from already overwhelmed landfills, composting is an effective way of minimizing the miasma of greenhouse gases that the rotting waste would otherwise emit, meanwhile building nutrient-rich soil to use in the garden.

So I built a compost pile by heaving four straw bales in a square, tossing in some horse manure and a thick layer of fresh kitchen scraps. Then I added another layer of straw, sprinkled the heap with water, and let the decomposition begin. But it wasn't until I bought my beautiful prized red worms—the Rumpelstiltskins of the soil who spin rotting lettuce into black gold—that the pile was complete. I fed them every few days—another layer of rotting fruit and vegetable parts, all rinds, pith, pits and seeds, coffee grounds, and eggshells—and kept the pile moist, careful not to add any meat, whole eggs, dairy products, or grease, so as not to attract attention from unruly rodents and other vermin.

Red worms are the lowest-maintenance and highest-producing organisms with whom I've ever had the pleasure of sharing time and soil. I highly recommend inviting some to your own garden.

Water

The land I was tending has an irrigation ditch that runs straight through the property. At the beginning of each farming season, the ditch always starts as a dry, muddy bed. Then the ditch master opens the gates and allows melt-off from the winter snowpack to flow out and down, through all the veins and arteries in the valley, to give the land a drink. After about a month of running thick and brown with sediment, the water trickles clear and calm. And by the end of summer, the streamlet is abundant with vines and bushes growing up along either side, a resting nook for an occasional mallard or thirsty heron.

The watering hole is a special place. Every morning I'd cross the rickety railroad tie bridge to turn on the spigot and listen to the water rush through the irrigation pipes. From the west bank, I'd gaze into the waters, thick with long, green algae, and sneak a raspberry from one of the nearby volunteer raspberry bushes, swatting at the hordes of mosquitoes, who'd been waiting silently for their next meal. Then I'd trek back over to the raised beds and begin to administer the water.

By July, the plants were in full growth, each bed abundant with expansive foliage and fruit, blossoms, and more than an occasional bind or pigweed. I took great care to greet each plant, watering it just enough to saturate the roots but not drown them.

Each plant variety has a different thirst. Some, like lettuces and a few herbs, have shallow roots systems and require an abundant but shallow drink. Others, like root vegetables, and tomatoes, whose root systems often wander feet down into the earth, like a long, deep drink, thereafter to dry out again between waterings. And for the squash, sunflowers, and raspberries, which I couldn't reach with the sprinkler hose, I'd fill and haul buckets to ensure that no plant went thirsty. Not the most energy-efficient irrigation method, but a sure way to become versed in the labor of love.

This is how I learned the personality of the garden. Every farm morning became a ritual to regard the sacredness of my homeland and its inhabitants: the robins who hopped from patch to patch of pasture, cocking their heads to listen, then bobbing down to yank out a worm or bug for breakfast; a gaggle of nuthatches who would startle and fly into the netted fence over and over again, until they'd finally launch themselves over the threshold as I approached; the dragonflies who would skim the water and land on a sunflower leaf to observe the sun's movement; and the precious plants who put all of their energy into bursting forth with fruit. In a garden, every nuance is alive with beauty.

ACCORDING TO THE USDA (U.S. DEPARTMENT OF AGRICULTURE), AGRICULture accounts for over 80 percent of the annual freshwater consumption in the United States. But freshwater supplies are rapidly diminishing due to pollution, climate change, and population increase. Meanwhile, multinational companies are staking their claim to global water supplies. The *Toronto Globe and Mail*, one of Canada's leading news sources, reports that water is quickly becoming a globalized corporate industry.

In July 2010, the United Nations declared access to clean potable water a basic human right. It seems only natural that the oceans, rivers, and streams be granted the same right. But as the water wars wage on, and the supplies continue to diminish, the future of clean water for *anyone* hangs in the balance. Historically, water has been an inexpensive resource, a given, for humans, agriculturally inclined and otherwise. This artificially low cost has encouraged all human water users, including irrigators, to use more water than is necessary, a habit that is going to have to be broken. And soon.

In order for agriculture to be sustainable through times of water scarcity, farmers will be forced to raise the productivity of their water sources. By applying conservative modes of irrigation, while considering soil type, temperature, and humidity of their given region, agriculturalists are on the forefront of restructuring consumptive water use. But we, the consumers, are accountable for much of the current freshwater predicament, as our food choices dictate what kinds of crops are being grown. And the thirstiest crop of all, which many Americans tend to include in almost every meal, is meat.

Livestock

Researchers have calculated that it requires 2,600 gallons of water to produce a single serving of steak. By comparison, it takes six gallons to produce a head of lettuce. Put another way, it takes ten times more water to produce meat protein than it does to produce grain protein. A statistic worth considering next time you order a hamburger.

Growing animals for human consumption has been historically common for thousands of years. There are many instances where livestock—including horses, cattle, goats, sheep, and chickens, just to name a few—can be beneficial to the land when husbanded correctly, responsibly, and in moderation. A healthy carrying capacity, or livestock-to-land ratio, and proper rotational grazing, wherein the pasture has the chance to regenerate before contending with livestock grazing on it, can add to the land's vitality. Most livestock manure is a rich amendment that can add valuable nutrients back into the soil.

But the introduction of factory farming has completely changed the relationship between humans, livestock, and land. Confined animal feeding operations, also known as CAFOs, are the ultimate expression of the industrialization of Nature. They are designed for highest possible production while using the smallest amount of resources—space, money, labor, and attention—which has resulted in living conditions similar to that of concentration camps. CAFOs hold large numbers of animals, most commonly cows, hogs, turkeys, and chickens—sometimes hundreds of thousands—in captivity for their meat, eggs, or milk. And the animals are subjected to gruesome living conditions until they are either slaughtered or too sick to produce.

The livestock are confined to such tight spaces that they are often unable to move in any direction or lay down comfortably. Bulls are routinely branded, castrated, and dehorned without painkillers. Dairy cows (which under natural circumstances bond, much like humans, to their young) are impregnated and separated from their calves a week after birthing and then are subjected to such copious and invasive milking that their udders are rubbed raw, often causing painful infections. Chickens and turkeys, which are kept in such close quarters that they cannot spread their wings, are routinely debeaked without anesthesia to prevent them from pecking themselves or their neighbors.

Artificial means of maintaining the production and supposed "health" of the animals, such as steroids, growth hormones, and unnecessary antibiotics, are used with staggering frequency. Recent studies have concluded that American girls are undergoing puberty earlier than usual and attribute much of the early onset to eating factory-farmed meat and dairy products. When a person perpetually consumes meat tainted with excessive hormones, the hormones too are ingested, confusing the body's ability to regulate its natural hormonal pattern and thus triggering an abnormal onset of puberty.

Meanwhile, the U.S. Department of Agriculture reports that in 2009, over 29 million pounds of antibiotics were administered to slaughterhouse animals. That's 70 percent of the antibiotics used in America for the entire *year.* Health specialists are concerned about this recent development, suggesting that judicious use of antibiotics would be preferable to such gross overuse, so as not to promote the emergence of drug resistance.

But super bacteria and blossoming preteens are not the only negative side effects of factory farming. The factory farm animals, who are as sentient as any human, are neglected, mutilated, genetically manipulated, and subject to disease, chronic pain, and crippling due to their living circumstances. And when their number is drawn, they are often killed in gruesome and violent ways. Many remain conscious as their throats are slit, as they are plunged into scalding water to be defeathered, or while they are being skinned or hacked apart. Would you like fries with that?

The environmental effects of factory farms are equally as disturbing. Confining so many animals in one area creates unmanageable amounts of waste, not only creating opportunity for disease and foodborne illness to fester, but also having a devastating effect on the surrounding air, water, and soil. The nearby ecosystems simply cannot sustain such overloads of excrement. And the spiritual costs of factory farming are immeasurable.

A healthy farm system, including livestock, is designed to work *with* the natural ecosystem, not against it. One such operation is James Ranch, in the Animas River Valley. Here they raise grass-fed and -finished beef cattle and Jersey milk cows who graze on four hundred acres of high-altitude, irrigated pastures with a fresh supply of crisp, clean river water and high-mountain air. The James family has been practicing sustainable agriculture in and around the Animas Valley of southwestern Colorado for upward of fifty years. Their multigenerational business enterprise also includes a sapling nursery and an organic vegetable and flower garden. Each element of their operation is integral to the health of the others, and they all work together to maintain the overall health of their land *and* their family.

Crop and grazing rotation promote healthy, balanced soil, which provides a stable base to support plant roots. Manageable amounts of animal waste fertilize the soil, without polluting it. The James family has built a sustainable farming ecosystem that works within their surrounding habitat.

Harvest

Harvest time comes in waves. As one sows seeds continuously throughout the season, so one harvests. Each day when you arrive at the garden, and pick up a bushel basket to relieve the plants of their heavy cargo, you cannot predict all the treasures you'll take home.

By September, almost all the plants had unfolded into maturation. The sugar snap pea vines had grown to over six feet tall and had begun to wither with the cooler temperatures, but not before providing me and my husband, and a few close friends, with a bountiful supply of the sweet, green, crunchy pea pods. I'd harvested fresh rosemary, French lavender, striped sage, cilantro, and flat parsley to enhance all of our home-prepared meals, and enough basil for batch after batch of pesto and caprese salad. I had snacked on lemon cucumbers and green beans while pulling weeds and gathered carrots, beets, and turnips from the soil. Some of the root vegetables had split, or grown around a rock or piece of straw, and looked like mandrake, those impish root vegetables that come alive when no human is looking. And I wondered what mischief these vegetables were up to when I wasn't around, taming the garden.

Few things are more satisfying than eating a ripe tomato off the vine, experiencing it at its peak of ripeness. It is a fully sensory experience. Your eyes scan the voluptuous foliage, looking for ripe fruit to pluck from the plant. You pull off your gardening glove and glide your fingers through the leaves until your hand meets the supple and soft fruit. Judging a tomato's ripeness is an art. Most tomatoes will still be firm to the touch but give, just slightly, under your touch. The delicate flesh of a ripe heirloom tomato curves under your grasp. You take it from the plant and bring it to your nose, inhaling deeply; the scent of earth, water, and seed that have alchemized into this edible gift. Then you take a bite. Your teeth puncture the skin and juice gushes into your mouth and down your chin. A blend of sweet and savory, with a hint of tartness and a whiff of truffle. This is a *real* tomato.

Not the cardboard variety that most often adorn supermarket shelves. The flavorless—and often nutritionless—tomatoes and other produce that we generally find at the big-box grocery stores of today are products of industrial farming methods, which employ huge amounts of chemicals (mainly in the form of insecticides, herbicides, and fertilizers), water, and energy and are responsible for massive soil erosion, wastelandification, loss of soil fertility, and pollution of the environment.

In its beginning, industrial agriculture was used in America as a method of providing cheap and plentiful food to the masses. It was considered a success because from 1930 to 2000 the output of conventionally farmed food steadily rose 2 percent annually, causing food prices paid by consumers to decrease. But you get what you pay for. Food raised by conventional methods decreased in quality. And while their organic counterparts have not yet been proven to have better nutritional value, they have been grown in a manner that is more conducive to environmental sustainability.

Sustainable agriculture cannot rely solely on organic farming practices to save us from the damage caused by human eating habits. But organic does appear to be a better model than "conventional farming." Organic farming methods combine scientific knowledge and traditional farming practices, drawing heavily upon the natural world and its processes for inspiration. The International Federation for Organic Agriculture Movements defines *organic agriculture* as "a production system that sustains the health of soils, ecosystems and people. It relies on ecological processes, biodiversity and cycles adapted to local conditions, rather than the use of inputs with adverse effects. Organic agriculture combines tradition, innovation and science to benefit the shared environment and promote fair relationships and a good quality of life for all involved."[6] Organic farming is a more holistic approach to agriculture wherein crop rotation, green manure, building healthy soils with compost and other natural amendments, and biological pest control are but a few of the techniques administered to maintain the overall health of organic farms.

Let's compare apples to apples. In recent years, many studies have been conducted to assess the difference in quality between organically and "conventionally" grown apples. While scientists haven't *yet* proven the nutritive difference between organically and conventionally grown apples (some studies show a vast difference, others show very little), the organic apple was raised conscientiously, while the conventional apple was not. *And* the organic apple has not been bathed in hazardous chemicals before reaching your mouth—and likely the conventional apple has.

And a simple rinse and scrub cannot rid conventional produce of the toxic residues from chemical herbicides, pesticides, and fertilizers, which have been scientifically linked to a cornucopia of illnesses such as autism, ADHD (in children), Parkinson's disease, and various types of cancer. Conventional agriculture is the costumed evil stepmother attempting to deliver poisonous produce to our door.

According to the United Nations Food and Agriculture Organization (FAO), current food production can sustain the global food needs for the 8 billion people forecasted to live on this planet by 2030.[7] This projection accounts for the likely increase in meat production and does not rely on the addition of genetically modified crops. It does, however, depend on the conventional methods of farming. But several field studies in the United States and the United Kingdom have revealed that organic farming practices can yield comparable amounts of commodity crops (such as corn, soybeans, and wheat) to industrial farms. And while the organic movement has gained great momentum in the last few decades, conventional farming still produces the vast majority of foods consumed in the United States.

A common objection to organic food is that it is more expensive than the industrially farmed variety. But most consumers don't realize how much we are already paying for conventionally grown food before it even reaches supermarket shelves. Our tax dollars subsidize the petroleum used in growing, processing, and shipping these products. We also pay significant monetary subsidies, and incalculable environmental and health costs, to maintain conventional farming. So the "cheap food," at least in the sense of being less financially and environmentally expensive to produce, is actually the organic variety.

Community

If everyone in the United States ate just *one* meal per week that was comprised solely of locally and organically produced foods, our country's oil intake would be reduced by 1.1 million barrels per *week*.[8] But the joys of buying local produce reach far beyond the opportunity for self-congratulation.

Have you ever been to a farmers market? They are extraordinary places that provide the onlooker with an accurate cross-section of the community. Permit me to take you on a tour of my local farmers market.

With mug of tea in hand, I walk into the bank parking lot where the Saturday morning farmers market is held April through October. The growers, artists, educators, baristas, and musicians in attendance take their duties very seriously and show up to share their gifts rain or shine, in spite of wind and freezing temperatures.

On the southeastern corner, a booth that specializes in vegan raw food is where I like to begin. I ask one of the smiling young beauties with dreadlocks, perfect skin, and hairy legs for a fresh shot of wheat grass, and she hops on the bicycle juice press to squeeze me a glass of the green elixir. Kids of all ages run around with sticky fingers from their fresh pastries, as their mothers keep a watchful eye, sipping coffee and hungrily catching up on adult conversation.

Depending on the time of year, the orchard man, who has few teeth but a lovely smile, tells me about his fresh selection of peaches, plums, and apricots. In the fall he displays a variety of apples and offers me a sip of freshly pressed cider. The sweet and tart tang charms my taste buds and I have no choice but to buy a gallon. I walk on and see the Hula-Hoop lady, who wanders about, her colorful handmade Hula-Hoops slung over one shoulder as she peddles her wares to tourists and locals. Beekeepers sell honeycomb, amber-hued honey in jars, honey sticks, bags of fresh pollen, and star-, bee-, and bear-shaped candles. And the men who sell potted herbs and the compost lady with her bags of compost-ready red worms (yes, this is where I procured the darling red worms for my compost pile) are in attendance.

Then comes booth after booth of growers who've lined up, stacked, and prepped their harvest. Carrots, radishes, beets. Braising and salad greens, spicy salad mix, kale, bok choy. Bell, jalapeño, and banana peppers and stripy purple and white eggplants. Potatoes, onions, garlic, and fresh herbs. The heirloom tomato folks in their horseshoe-shaped booth point to various red plastic bins, each containing a different brilliant color of heritage tomato. Black truffle, early goliath, black krim, and brandywine—each tells a story of where it has been and all the flavors it has picked up in the soil along the way.

Coffee, tea, and lemonade stands dot the perimeter of the market, and artisan birdhouse carpenters and seamstresses of boutique aprons and bonnets explain their creations. One can peruse sheep wool, yarn, and lanolin soaps or ground blue and yellow corn from the Ute Indian woman's table. Bakers stand at the ready with that morning's supply of raspberry tarts, cheese Danishes, and assorted muffins. Their competition across the aisle offers long baguettes; scones; bags of sweet nutty granola; green chile, plain, and chocolate croissants—a feast of confectionary goodies that feature locally and organically furnished ingredients. In the nearby music tent a vocalist strums and sings to enhance the shopping experience. But we're not shopping; we're sharing the joy of our community and feasting on the delights of our local foodshed.

Dr. Nabhan points out that the "environmental movement" has always taken a stand against things like fossil fuel development and corporate agriculture, which are damaging Earth's wild places and the biodiversity therein. And as a result of its forbidding response to what is happening in the world, many have turned away from said movement as a viable way of making a positive contribution to the greater good. But he upholds that supporting local foods provides vast opportunity to take a more holistic, inclusive approach to saving the planet. And though supporting local foods is but a branch of the environmental movement, its impact is far reaching.

"Ecological thinking is not simply against the bad things that happen in the world," he says. By tasting the local foods, and investing in the bioregional diversity of foods that reduce our carbon footprint, we have sensory feedback mechanisms telling us that eating this way is not only good for the Earth and our bodies, but it's giving us more pleasure and deepening our connection to place and culture than what the big companies can feed us. "We're choosing to participate in something that is good, healthy for us and for the land, that gives us deep pleasure and spiritual connection to the land," says Nabhan.[9]

"Whether that food is of high nutritional quality or not is almost beside the point," Nabhan contends, "because the freshness that we get from a farmer's market, the connection that we feel with our neighbors and the land that we live on, makes food less of an abstraction and something that we're part of."

Collaboration

"A food culture is not something that gets sold to people. It arises out of a place, a soil, a climate, a history, a temperament, a collective sense of belonging," says Barbara Kingsolver in her book, *Animal, Vegetable, Miracle.*[10] Over the last few generations, our society has been duped into eating habits that are killing us and our environment. Thinking back, have we ever had a collective food culture in this country? Traditional Native Americans still seem to have some semblance of food culture, though they too have suffered the ill effects of mass marketing and food production.

As we progress into the twenty-first century, we have an opportunity to modernize—or perhaps traditionalize—our food culture into one that does not rely on fad diets or a profit-driven food industry for inspiration. But it will have to be a collaborative effort to (re)discover the cherished elements of food culture that other countries have taken millennia to create: what foods comfort and nourish us? Do they grow anywhere nearby? Our poor eating habits have become tradition, and if tradition is the bearer of food culture, we're in big trouble.

But something is beginning to happen in the West, which could be a harbinger of hope. People from diverse backgrounds—Native Americans, scientists, economists, ecologists, psychologists, parents, farmers, all who eat—are beginning to realize that we have something in common: that we share common water and food systems.

As Dr. Nabhan puts it, "People [are] collaborating beyond the narrow interests of individual land owners to allow whole systems to function cohesively across boundaries."[11] People from all sectors are beginning to pool their resources, be they financial, physical, or cognitive, in order to heal damaged ecosystems. They are making any contributions they can toward the common good, which is the direction humans will have to collectively migrate to if we have any chance of survival as a species. Humans are moving into an age where if we wish to survive, we can no longer remain ignorant and apathetic about any of our choices, but especially those surrounding food. Every bite that we take leaves a tangible, indelible footprint on our global ecosystem.

As long as we only express concern for the stewardship of *our* own land and *our* own pantries, the health of our communities, and in turn ourselves, is in peril. But if we can pool our resources on a local, regional, and eventually a global scale, there is still potential for human survival. By doing so, we can heal the broken systems that have been degrading our spirits, souls, bodies, and ecosystems, and renew their collective integrity.

But as a relatively new country, America takes itself too seriously. As time progresses and our traditions evolve, we would do well to revive our collective sense of humor. Dr. Nabhan offers, "Whether you're an Evangelical Christian or an evangelical environmentalist, as long as you believe that you're above being vulnerable and flawed and that you're above making mistakes, we will continue to make mistakes. And the land's wisdom will not be transferred into human wisdom."[12]

Humans, especially in the West, have a tendency to try to "fix" Nature to increase its productivity and hospitality. The sooner we realize that the environment is an ever-evolving entity, to which *we* must adapt, and not the other way around, the more likely we are to flourish within our given homeland. Humans can only begin to understand the complexities to which Nature is inclined, and though we have made many intellectual breakthroughs, our intuitive understanding of landscape has diminished. Perhaps by allowing our intuition to guide our intellect, we can begin to overcome the urge to dominate the landscape.

Humans are but environmental stewards, a mere thread in the tapestry of our ecosystems, and the elements will remind us of our folly should we continue to ignore this wisdom. If we can remember that we are a *part* of Nature, and act and eat accordingly, then we can once again find the balance needed to work toward a mutually nourishing and sustainable future. We have a lot of work to do, but should we bring our humility, humor, and a *healthy* appetite to the table, the results will be downright delicious.

Notes

1. Wendell Berry, *Bringing It to the Table* (Berkeley, CA: Counterpoint, 2009), 31.
2. Ibid., 92.
3. Masanobu Fukuoka, *The One Straw Revolution* (New York: *The New York Review of Books*, 2009), xxvii.
4. Center for Food Safety, "Genetically Engineered Crops," accessed May 22, 2012, http://www.centerforfoodsafety.org/campaign/genetically-engineered-food/crops/.
5. Gary Paul Nabhan, interview by Jack Loeffler, 2010, transcript in author's possession.
6. International Federation of Organic Agriculture Movements, "Definition of Organic Agriculture," accessed May 22, 2012, http://www.ifoam.org/growing_organic/definitions/doa/index.html.
7. United Nations, *World Population Monitoring 2001: Population, Environment, and Development* (New York: United Nations, 2001), 16, http://www.un.org/esa/population/publications/wpm/wpm2001.pdf.
8. Barbara Kingsolver, with Steven L. Hopp and Camille Kingsolver, *Animal, Vegetable, Miracle* (New York: HarperCollins, 2007), 5.
9. Gary Paul Nabhan, interview by Jack Loeffler, 2010, transcript in author's possession.
10. Kingsolver, *Animal, Vegetable, Miracle*, 17.
11. Gary Paul Nabhan, interview by Jack Loeffler, 2010, transcript in author's possession.
12. Ibid.

Gary Paul Nabhan, photo by Jack Loeffler

Gary Paul Nabhan

Gary Paul Nabhan is an agricultural ecologist, ethnobotanist, and writer whose work has focused primarily on the plants and cultures of the desert Southwest. He is considered a pioneer in the local food movement and the heirloom seed saving movement.

A first-generation Lebanese American, Nabhan has served as director of science at the Arizona–Sonora Desert Museum and cofounded Native Seeds/SEARCH, a nonprofit conservation organization that works to preserve indigenous southwestern agricultural plants as well as knowledge of their uses. Nabhan was the founding director of the Center for Sustainable Environments at Northern Arizona University. In 2008 he joined the University of Arizona faculty as a research social scientist with the Southwest Center, where he now serves as the Kellogg endowed chair in southwestern borderlands food and water security. Nabhan won the John Burroughs Medal for distinguished natural history writing. He was also the recipient of a MacArthur Fellowship.

The unifying theme of Nabhan's work is how to avert the impoverishment and endangerment of ecological and cultural relationships while celebrating the traditional ecological knowledge of the agrarian communities. He has played a catalytic role in the multicultural, collaborative conservation movement, being one of the coauthors of its populist manifesto, Finding the Radical Center. Nabhan was among the first creative nonfiction writers to link the loss of biodiversity to the loss of cultural diversity. Nabhan now attempts to restore nectar corridors for pollinators in binational watersheds around his home in Patagonia, Arizona, which he calls the "pollinator diversity capital of the United States."

Nabhan farms a diverse set of heirloom fruit and nut varieties from the Spanish Mission era and from the Middle Eastern homelands of his ancestors, as well as heritage grains and beans adapted to arid climates. He is a champion of water harvesting, which he implements in his own orchard and gardens. Dr. Nabhan is regarded as one of the most important environmental activists of our time.

Restorying the Land

(The following is excerpted from four interviews conducted over a period of fourteen years by Jack Loeffler.)

GPN: In my book, *Desert Legends*, one of the key notions is that to restore the land, we have to restory it. And I think this is a missing element of most restoration work. Whether the stories come to us through poetry or oral history, odyssey and other narratives, or whether they come to us through some combination of music and performance, these ritualized parables tell us who we are and how we should relate to the land in a way that underscores the primacy of that relationship. We re-member the land, and the land community makes us a member once more, not just a "head way," but a "heart way" and a "gut way." Its story is in a very real sense our story, and people who forget the story of their relationship to the land are the people who have become most vulnerable on this planet.

In his book *Who Owns the West*, Bill Kittredge has said, "It is not simply that individuals can forget their story, their autobiography, but whole cultures are doing it now." He feels that in the American West, the Anglo-American culture has forgotten its story. Or perhaps it determined that its original story of the frontier epic is now obsolete, it's not guiding us anymore, and that we need to find a deeper story, something that can freshly direct us and inspire us for a longer period of time. How can we find those very stories which restore our ancient connections to the land and build upon them so that we have a bedrock solid sense of place?

For my adult entire life, I've been interested in the endangerment of biological and cultural relationships. We know we are losing species, seedstocks, and breeds all around us—that is the loss of biodiversity. But we are also losing languages, and with them, the erosion of traditional ecological knowledge is also occurring at an alarming rate. We're witnessing an extinction spasm of linguistically encoded cultural information about the natural world. It may be the most serious extinction threat that we're facing today. There were once twelve thousand or so languages in the world. Within the last eight thousand to nine thousand years we've lost six thousand of those, probably most of those within the last three hundred years, and we're expected to lose another half of the extant languages in the world within the next decade. That is to say that in North America half the Native American languages that currently persist won't have even two speakers in the same village who can talk to one another in that language within the next generation . . . unless we restore those relationships.

I don't think it's any coincidence that we're going through an extinction spasm of biodiversity and an extinction spasm of cultural diversity at the same time. That comes back to your question about those cultural stories about how to live well on the land without abusing its resources. That's not to say that every indigenous culture has lived in some kind of homeostatic balance with the land. We both know of cases where reservation lands have been overgrazed, little valleys have been overhunted, people have depleted fuel wood resources from places in Indian country where the fuel wood grows very slowly.

Nevertheless, those stories are there to remind people of the long-term consequences of doing such damage, so that their behavior can adjust and their behavior can remember timespans longer than the human lifespan and how some trees live longer than we do. So I think when we lose those languages we're losing some of the ethical cautions, the ways that we patrol ourselves, the ways that we remind ourselves that there are consequences of mean-spirited action or ignorant action in relation to our fellow constituents of this natural community.

The Yaqui, Huichol, and Tarahumara people probably have more of their languages in place and their cultural rituals in place than most cultures in northern Mexico. Again, here's the relationship between diversity and biodiversity right in front of us. The Sierra Madre Occidental—their Mother Mountains—is the richest landscape in biodiversity north of the tropics anywhere in the New World. Linguistically it's not just the Tarahumara and Yaqui and the Huichol, but Guarijio and the Cora and the Tepehuan and Mountain Pima and the Mayo and other cultures—a very high level of linguistic diversity associated with those Sierra. And yet they are now reeling from impacts like nothing ever seen previously, completely unprecedented scales of logging, and road building, and climate change going on in the mountains of Durango and Chihuahua and Sinaloa and Sonora.

Sadly, new roads are making it possible not merely for logging and milling but also the spread of invasive species and the spread of livestock in densities that are unprecedented for that area. Livestock has been in the area three hundred to four hundred years, and yet the densities of herds are unprecedented. We have a very dramatic kind of clearcutting happening, mostly by American-funded companies that are taking very ancient trees and pulping them for making our Kleenex and toilet paper. We also have the understories of these forests being grazed out by narcos posing as cowboys. That will change the fire frequencies and the availability of the *quelites*, the wild greens the Tarahumara and the Yaqui use. They are also likely to lose many of their medicinal and ceremonial plants. So this erosion of habitat is clearly related to the displacement of people, language loss, and cultural loss and the impoverishment of diets, rites, and pharmacopeias. These are all related. If their habitat is being eroded upon or encroached upon, there will be dramatic changes in the next thirty or forty years.

Human rights activists and conservation biologists have joined together to stop a major World Bank logging project. But the World Bank plans for the Sierra Madre attracted the attention of so many private logging firms that the forest may not be any better off. Activists could patrol the World Bank, but it's much harder to patrol 120 private industries.

JL: What you're saying right now is bringing to mind the Sierra Madre Occidental role as both a biotic and cultural seed bank. Within the context of the Mexican people themselves, are there any real attempts to try to preserve the Sierra Madre Occidental, its biota, and its cultures?

GPN: Absolutely. Look at the example of the late Edwin Bustillos, one of twelve people around the world to receive a Goldman Foundation Award one year for his extraordinary commitment to forest conservation and forest cultures. His life was threatened at least five times. He continued to organize collectives and cooperatives to set up their own forest reserves in order to keep the entire landscape from being logged. He was very important in slowing the encroaching of drug lords on indigenous peoples in the Sierra Madre. I think we need a thousand people like Edwin Bustillos in the Sierra rather than just three or four. [He died in 2003.] So there are exceptional people like Edwin who have provided a model to inspire local community members. If he can stand up against such pressures, why can't local people take charge of the fate of their own landscape and repel outsiders who don't wish to honor unique and cultural communities?

There are other people working in the Sierras. Native Seeds/SEARCH has had a reforestation and orchard project. Randy Gingrich has done excellent monitoring of forest encroachments and getting injunctions against illegal land takings and logging in Tarahumara country. On top of that, he provided backup support to Edwin before he died. But it is really at a point where we need thousands of people with eyes and ears and extraordinary antennae out walking and hiking through the Sierra, kayaking down its rivers, to find where all these encroachments are going on so that we can just keep track of them. The landscape is changing very rapidly. It is the most diverse left in North America—a true global hotspot of biodiversity and cultural diversity.

Because we've starved the delta to death and starved the upper Gulf of California to death, we now don't get the full benefit of that wildlife corridor even here in the United States. The Colorado River itself was once part of this corridor, but without the riparian vegetation along it and the productivity connected to it, it has diminished value relative to what it had over a century ago. If you ever want to imagine a dramatic difference between how a river was managed historically and how it is managed today, look at accounts of it that predate Aldo Leopold. He was among the last to witness what the Colorado delta and the lower Colorado were once like.

A hero of mine, Robert Forbes, was the first dean of the College of Agriculture at the University of Arizona. In those days, a university dean felt compelled to know the terrain, so he took a trip down the Colorado several times in the 1890s and early 1900s from where Fort Mojave is today down to the delta. He recorded rich details about how at that time, native farmers practiced floodwater recession agriculture, where people would wait to plant their summer crops after the spring snowmelt increased the flows in the Colorado. They would go out onto the floodplains' muddy bottoms, sometimes spitting seeds out of their mouths. They were broadcasting panic grass and other crops unique to the Colorado River system as well as other ones that had been exchanged over the millennia with Mexican Indian farmers. They had an oasis agricultural system very much like what we see on the Nile along its most remote reaches today. That "watershed agriculture" was uniquely adapted to the Colorado River Watershed's heat, salinity, soil fertility, growing season length, and was absolutely unique in terms of anything else in North America. But by the midteens, that mix of wildness and culture was disrupted. When Aldo Leopold and his brother Carl canoed down the delta, they saw only the remnants of that system. Cucupa Indian agriculture was already gone and growing back in native and exotic plants. The wonderful mosaic of cultural landscapes intermixed with wild landscapes was already lost by the time Aldo Leopold reached the delta.

So when we talk about watershed thinking we're not only talking about reflecting on geophysical and ecological attributes of a watershed but on the cultural adaptations to it as well. Those three elements should guide any residency or any production in an ecosystem, and yet we've broken the connections between those physical attributes, the agroecological attributes and the cultural attributes. Now the people who farm in the Yuma Valley or the Imperial Valley do not do so with any cognizance of the history of that watershed or where their water is coming from, for that matter, or what animals, plants, and habitats they may be starving by the very way they intensively farm using water and energy from elsewhere in the watershed.

JL: One obvious geographic irony is that when you get up into the northern ends of it, the headwaters of all these different places including the Green, the Colorado, and the San Juan rivers and various others, you have more of a feeling of ruralness, almost as though you were still in the nineteenth century. Then as you work downstream it grows evermore impoverished aquatically until you get to the Hoover Dam, at which point landscape impoverishment accelerates with increasing industrialization and urban sprawl.

GPN: Right. From the river trip where you and I were joined by Gary Snyder and others on the San Juan in 1999, I've done three chapters linking that to the entire Colorado River–Sea of Cortez ecosystem for a book called *Coming Home to Eat.* Basically I used our river trip in the San Juan River wildlands as a state-of-the-hinterland report. I then look at the agriculture in southern Arizona as how this foodshed and watershed feeds us. And of course those bottomlands are dependent upon the soil washed down from the headwaters, not just the water. Why we have the great fertile valleys of the Imperial and Yuma is because of the tremendous fertility that the headwaters have shared with it. Next I look at the fisheries in the Sea of Cortez that have been depleted in part due to exploitation of shrimp and other shellfish by trawlers but have also been diminished because of being robbed of all their freshwater and fertility flows. So if we think of the Colorado as a foodshed as well as a watershed, every bit of food produced down in Yuma and Mexicali is dependent upon hundreds of years of water and nutrient flows from the headwaters. You can't get that kind of productivity if you don't have a well-functioning river. So we're living off the capital that the Colorado River headwaters produced over millennia and we're rapidly depleting it.

One other really interesting aspect of the Colorado River is that we have a tremendous survival of Native American languages in the headwaters and hinterlands regions of the watershed, but they're under great stress from the Hoover Dam down. Let's talk about that for a second. The Colorado Plateau itself has more distinct languages and dialects of languages than any remaining ecoregion of North America beyond Mexico.

At the same time we have more language speakers of indigenous dialects here in the Colorado Plateau than all the rest of North America combined. About 53 percent of the Native-language speakers residing anywhere in the U.S. reside on the Colorado Plateau. South of Hoover Dam all of those language groups are under severe stress: Quechan, the Maricopa, the Hia 'C-ed O'odham, the Mojave, the Chemehuevi, the Cucupa, and there is a smaller and smaller percentage of the children learning those Native American languages. So the tremendous amount of cultural knowledge that is encoded in indigenous languages is being lost to us at a very rapid rate downstream from Hoover Dam.

JL: Apropos of that, I've been thinking that within a particular language is borne all of this knowledge about to how to think of the watershed in a proper fashion. Many of those people have traditionally seen themselves as part of the land itself rather than apart from it. Subsequently if that could coalesce within the context of today's greater purview of the way water in the Colorado River Watershed is used, it might lead ultimately to a federal act where, thinking off the top of my head, an Ecosystem Protection Act or Watershed Protection Act could be created that was commensurate with the protection of the linguistic phyla—I recall that there are three linguistic phyla present on the Colorado Plateau.

GPN: Yes, three.

JL: And those three linguistic phyla themselves contain the knowledge of how to comport culturally within the watershed.

GPN: Yes, perhaps twenty-two different languages and their dialects. Different dialects of Navajo, for example; and perhaps in Hopi, Tewa, Paiute, Walapai, Yavapai, Havasupai, and then Shoshone, Chemehuevi. The Shoshonian groups used to venture into the watershed of course. And then Cocopa and Quechan on further south. But the remarkable thing is that their oral histories encode a tremendous amount of knowledge of environmental change. For instance, in Canyon de Chelly there are Navajo names for streams that imply that they were perennial at the time that the soldiers first came into that area. They're no longer perennial streams. There are hundreds and hundreds of place names from the Colorado River uplands that can tell us about the environmental changes that have occurred over the last century, century and a half, and environmental historians have hardly used that kind of information to give us a depth of the dynamics of this ecosystem.

The same thing is true with our potential to recover it. There's still a tremendous wealth of Hopi oral history about condors in Arizona. We forget that the "California" condor was not exclusively a California bird. Well into the 1870s and 1880s there were sightings of condors, by Hopi folks going down into the Grand Canyon by way of the Little Colorado/Colorado River confluence to obtain salt. Those kinds of oral history elements encoded in Hopi could also help better situate the recovery plans for condors and other species that were on the verge of extinction. Native knowledge could now help them.

One interesting thing that you brought up earlier is the competition for water among urban areas, the tribes, and agricultural areas. I think that one of the greatest fallacies that has been put forth in the media is that agriculture is more wasteful of water than cities. The waste levels are much higher for urban uses than for agricultural uses, where wildlife often directly benefit from slightly leaky acequias and such. Of course, urbanization offers none of the environmental amenities in terms of open space or cleaning the air or wildlife habitat that healthy agriculture offers us with its ecosystem services. So cities are forcing the retirement of agriculture to get its water all across the Sunbelt. The result? Since 1982, one-quarter of all farmland loss in the entire U.S. has been in just four states: Arizona, New Mexico, California, and Texas. Six million acres of food-producing capacity lost for condos! Originally the Central Arizona Project (CAP) was planned to help Arizona farmers, and now most of its transfer of waters is clearly being used to fuel urban growth rather than promote a healthier agricultural system.

JL: I'm currently producing a one-hour-long radio documentary focusing on the life of Aldo Leopold who died before his great classic *A Sand County Almanac* was published in 1948. One of the things I wanted to ask you to ponder is how to extrapolate how Aldo Leopold's thinking can factor into the next thirty or forty years of environmental thinking.

GPN: I believe that the way to move Leopold's vision along to the next step is by restoring our watersheds and foodsheds as the basis for redressing our relationship to this planet. We need to plan and act with our watershed and foodshed in mind, and now we need to bring those notions from being merely poetic and into the pragmatic. We know that Powell talked about a vision of the West that was based on the sovereignty watersheds. The entire Leopold family has had much to teach us about this.

Today I think it's just as important to see how those watersheds relate to these foodsheds and ultimately to our own food security. The "Bureau of Defamation" [Bureau of Reclamation] has done transwatershed transfers of stream flows to favor food producers or cities on one side of the watershed divide at the expense of people on the other side. But what I think we need to do now is see how that water and its highest use is not a use that goes to the highest bidder but to the highest diversity of trophic users—that fish, cottonwoods, Hispanic farmers, Native American herbalists are all engaged in that cascade of water and energy through foodshed/watershed. When I've done surveys of Arizonans about their willingness to see policy change to ensure that our watersheds and foodsheds function over the long haul, over 70 percent are in agreement that we should, at each watershed unit in the Southwest, decide how much land and water should be set aside into perpetuity to assure that we can achieve local food self-sufficiency. They want to do this without compromising the survival and recovery of other species.

The interesting thing there is that even though we now send water to the highest bidder—L.A., Phoenix, San Diego, or El Paso—most people on the land are in agreement that right now at this very moment we need to turn that process around and set aside in every watershed in the Southwest the amount of land and water to sustain our biotic community.

Some of our elder statesmen like Wallace Stegner have said that the two greatest contributions that America has given to the world are wilderness and jazz. And it's not that there weren't elements of those two impulses—of improvisational music and wilderness—in other cultures before the U.S. gained an influential status. But they coalesced here in a very, very particular way that allowed those values to be elevated. Jazz reached some of its greatest expressions in the United States, as did wilderness. And the seeds of the wilderness movement came from Aldo Leopold's fascination with and immersion in the Gila Wilderness not far from the Arizona–New Mexico border. When my children were small, I took them for their very first backpack into the Gila Wilderness when their backpacks literally hung down to their calves. And we ate wild potatoes and wild berries and fished for trout there in the Gila Wilderness. What is remarkable to me is that it is the kind of Southwest that was cherished not just by Anglo recreationists, but that it was one of the last refuges of the Apache people before they were subdued by the U.S. government and that there have been rich Hispanic communities on the edge of that wilderness area that have drawn strength from it for many, many decades.

So that we have in that area this incredible expression of three cultures' contributions to our sense of wilderness: the deep history that our Native American brothers and sisters provide us; the rich traditions of Hispanic cultures that remind us that people can be transplanted from another land and make a new land their place; and perhaps the capacity of northern Europeans to codify the values of wilderness into a formal designation that could then be shared with other people in other parts of the world.

JL: That brings up a question about those of us who have blown in like tumbleweeds over the last single lifetime like I did about fifty years ago. In other words, how does one establish or reestablish a sense of indigeneity?

GPN: Perhaps Aldo exemplifies this, for Leopold lived in three worlds really. He grew up in that Iowa ecotone between forest and prairies, and that ecotone of Nature inspired him even though there may not have been much of a cultural ecotone there. But then very early on as an adult he came out to the Southwest where he saw that cultural ecotone as well. He recognized the incredible creativity that comes from learning the different past indigeneity that Native Americans, Hispanics, and Anglos in the Southwest had in the preindustrial era. And then he had a third way of being in the world that perhaps was refined with his time at Yale and then as a professor at University of Wisconsin, where he continued to immerse himself in great literature, while writing what many of us consider to be some of the great land-based poetry of the twentieth century. He was to land-based writing what Hank Williams was in the same era to country music or to what Ferde Grofé was to creating an American classical music. All three of them were contemporaneous, in ways trying to do the same thing. There's not much difference between *I'm So Lonesome I Could Cry*, *The Song of the Gavilán*, and *The Grand Canyon Suite*. These works, like *Leaves of Grass*, are the highest artistic expressions of *rooting*.

The point is that each of them was desperately trying to psychically root themselves in a new land, and they knew that that happened through internalizing the song of that place and participating in that place and not leaving it an abstraction. A sense of place is not something that we can gain nationally. It's something that we gain through full bodily, mental, and spiritual participation, participation with all our senses, so the smells, tastes, textures, and aromas of the land permeate our whole being.

Leopold's primary unit of care and concern was the biotic community, including human cultures. And it's ironic that just as he died in 1948, the word *community* began to be replaced by *ecosystem* in ecology. Scientists then used a mechanistic term from systems engineering to replace a sociological term which implied that Nature is a communion, that all members of a biotic community have some cross-referential influence upon one another. Very soon after that, we really began to look at them as spokes in the wheel or temporary eddies of energy in the flow of physical forces across the land. And so ecology became much more mechanistic almost immediately after Leopold died. For that reason *A Sand County Almanac* wasn't read much in the early '50s and even the '60s. It wasn't until '68 or '69 that people rediscovered it. It had only sold a few thousand copies by that time but then sold more books in the five years after the first Earth Day in 1970 than it had sold in the previous twenty years. A quarter century after it was written, it became a bestseller. The same thing happened with *Moby Dick*! We Americans are slow to absorb messages from our most rooted wisdom keepers!

GPN: Many people think that my work with Native Seeds/SEARCH and Seed Savers Exchange is only about saving seed diversity, as if all we need is more diversified genetics in our corn and beans and squash. But perhaps we need the seeds to save certain sensibilities in American cultures more than the seeds need *us* to save *them*. What we're really doing is using seeds as a symbol, as a metaphor, a light to shine upon the need for structural diversity in our agriculture, not just small farms but all scales of farms nested in a diverse community. We need good Anglo farmers, good Hispanic farmers, good Hopi farmers, all who have different points of view on how to use the land and how to see things in the land that one culture alone cannot see. So I'm really not just focused anymore on the diversity of livestock breeds or seeds as if that alone can keep agriculture healthy. We really want that to be part of a suite of strategies that I think sanction, justify, and celebrate the cultural diversity that has generated the great stewards and diversifiers of domesticated plants and animals and the habitats in which they flourish.

So when you think of Hopi corn, just don't think of multicolored corn. Think of how each Hopi village, each clan, has selected that corn differently over thousands of years and how they've interacted with the Tewa and Navajo and southern Paiute and perhaps even the prehistoric Hohokam and Anasazi, the prehistoric Puebloan people. And each of those cultures has selected corn for different characteristics: for drought tolerance, for deep rooting, for use in making hominy corn, for use in making piki bread, for use as a cornmeal for blessings. And without that whole suite of cultural viewpoints we would never have gotten to the point where today we can see thirty different races of Hopi corn still grown in an area about the size of Albuquerque.

JL: A concept that we're really trying to forward in this whole project is looking at commons as a vital artifact conceptually, and not only an artifact but an approach to redefining our relationship to habitat, to homeland, and subsequently we've taken a huge cue from that John Wesley Powell map that shows the watersheds as potential *commonwealths*. It would be impossible to culturally shift into a new paradigm overnight, but can you imagine how a sense of the commons could be factored in to a more diversified cultural point of view?

GPN: The most exciting thing happening in the West from my perspective is the collaborative conservation movement, where people of diverse backgrounds are saying, "We share this watershed in common and should comanage it. We share this foodshed in common and our co-ops nurture us." And that means people collaborating beyond the narrow interests of individual land owners to allow whole systems to function cohesively across boundaries. We're beginning to see what some people are calling now *landscape labeling* or branding of foods from a watershed or foodshed that are grown sustainably along with the fiber and the timber from there. Members of agrarian communities who come together to embrace the same stewardship principles will get a premium for their products, whether it's beef or vegetables or timber or fuel wood because urban consumers will be willing to invest more in that land stewardship and the ecosystem services functioning over entire landscape rather than just paying for the product like beef or cucumbers or firewood alone.

So what I see is that over the last twenty years we've seen incredible innovation in how people farm and ranch. Individual families have demonstrated means of sustainability in different habitats all over the West. And yet many of their neighbors have not had the capital, knowledge, skills, or public support to move in that same direction. So many of the key innovators—I think back to Peter Van Dresser in your area—were talking about these issues thirty years ago but never saw their neighbors embrace these same practices. Why? Because up until now there has been no economic incentive to get over the hurdles to restore land that's been badly damaged. To some extent, as long as we say, "I'm only concerned about the stewardship on my piece of private property," the health of the entire watershed will remain imperiled. But if we collectively embrace returning the entire watershed to its ecological health and mobilize financial as well as knowledge resources to do that, then there will be lasting benefits from the kind of innovations that we've seen over the last twenty years.

JL: Something that I'm contending with when I interview certain ranchers is the deeply embedded sense of the territorial imperative that exists.

GPN: Powell historians like Bill deBuys and Don Worster have pointed out a very interesting paradox about the West: that we emphasize in the myth of the West rugged individuality, and yet because of the Reclamation Act we actually have one of the most socialized landscapes in the West—at least in terms of water management—of any landscape in history. That the Bureau of Reclamation has basically socialized the availability of water, first to homestead farmers and then to corporations, and then a lot of that water is now being shunted to the cities. Phoenix would not exist if it were not for that Bureau of Reclamation mentality that's essentially government-managed socialism. There are very few rugged individuals in the West that are not getting subsidized water or subsidized grazing or subsidized road building to manage their little patch of land. And the sooner we admit that sense of interdependence rather than thinking that we're still in a society of independents, the better off we are.

JL: I have one last thought, Gary. You've probably done a fair amount of reading about future probabilities with regard to global warming and climate change in the Southwest. Do you have any sense of how that could work out?

GPN: When I hear the term *global climate change* I reframe it as region-by-region destabilization, with an increase in the uncertainty and unpredictability of each of these independent systems. We need to listen to people on the land that are already living with that uncertainty every day, as I tried to do with my friends Kurt and Kraig in the book *Chasing Chiles*. I don't think global climate change will be settled by a top-down solution from Kyoto, Copenhagen, or Washington. I spent the last year interviewing heirloom chili pepper farmers from the Yucatán Peninsula to the Río Arriba, northern New Mexico to St. Augustine, Florida, to Avery Island in Louisiana. And I trust the vernacular capacity to adapt to changing conditions and to find innovations emerge from the stuff in front of us. I don't think that top-down mandates from governments or from the scientific priesthood are going to get us where we need to be. So I'm back to believing that community-based knowledge is more dynamic than either governmental dogma or scientific dogma and will teach us how to adapt to the expectations of this always-dynamic landscape.

JL: Amen, brother Gary. Thanks.

William deBuys, photo by
Jack Loeffler

William deBuys

William deBuys is a celebrated author and a former professor of documentary studies. He serves as a private conservation consultant whose clients have included the Conservation Fund, the Nature Conservancy, the U.S. Forest Service, and the National Biological Survey. He is the author of several books and articles and the recipient of numerous fellowships and awards. His book *River of Traps* was honored as the *New York Times* Notable Book of the Year in 1990. Other awards have included the Robert Langsenkamp Award, the Pflueger Award, and the Watershed Steward Award. His first book, *Enchantment and Exploitation: The Life and Hard Times of a New Mexico Mountain Range*, which won a Southwest Book Award, combines the cultural and natural history of northern New Mexico. His second book, *River of Traps*, one of three finalists for the 1991 Pulitzer Prize, combines memoir, biography, and photography. DeBuys's third book, *Salt Dreams: Land and Water in Low-Down California*, is an environmental and social history of the land where the Colorado River comes to an end. He is also the author of *Seeing Things Whole: The Essential John Wesley Powell* and *The Walk*. His most recent book, *A Great Aridness*, addresses global warming and climate change in the North American Southwest and is one of the most important books of our time as it apprises the reader of the absolute uncertainty of the future of our species and many others as we inhabit a planet that has been unalterably changed by human presence. In the following pages, he reveals his take on our cultural approach to the array of hazards that we face in the imminent future.

Navigating the Rapids of the Future

(The following is excerpted from an interview conducted by Jack Loeffler on July 6, 2011.)

JL: Basically, Bill, the project we're working on now is called "Thinking Like a Watershed." Conceptually, it has its origins in that map rendered by John Wesley Powell in the nineteenth century that you gave me a copy of years ago. It's become part of my consciousness, actually. Two watersheds that we address are the Río Grande and the Colorado River. I know you've been doing a whole lot of research regarding global warming in the American Southwest. Could you give me an easy overview about your sense at this point about the potential probabilities or possibilities with regard to global warming and climate change for the two butterfly wings of the Colorado–Río Grande watersheds?

WD: That's a big question, Jack. I think one of the main things that climate change is going to be telling us is to do the things that we should have been doing anyway, like living within a sustainable water budget. Right now on the Colorado River, according to one of the main individuals who directs the use of the Central Arizona Project, the Lower Basin of the Colorado (south of Lee's Ferry) is operating at an annual deficit of between 1.2 and 1.3 million acre-feet. That's an enormous amount of water. The allocation for the Lower Basin is only 7.5 million acre-feet, so it's a big percentage of that. Basically the reason that's happening is that none of the states in the Lower Basin (California, Arizona, and Nevada) budgets its water consumption for evaporation, transmission losses, their share of the apportionment that goes to Mexico for our treaty obligation, things like that. They have been expecting, and have been getting away with taking, surplus Colorado River water coming from the bounty of the weather or resulting from the failure of the Upper Basin to use all of its allocation.

Well, in recent years the Upper Basin has not been sending down much in the way of leftovers—it has been using more and more of its allocation—and, aside from a big-water year in the northern Rockies in 2011, the weather has not been very bountiful. Inflows to Lakes Mead and Powell, the big reservoirs on the Colorado, have been slowing, and the levels of the reservoirs have been dropping and dropping and dropping. Although 2011 provided a kind of temporary stay of execution, we've come very close to having to take some serious measures to deal with these annual deficits. In fact, in the last couple of years we've had pretty good winters with pretty good snowfall in the Colorado Basin, but the runoff has not been proportionately as good. What's going on? It's probably climate change. We've got warmer temperatures, and that means more evaporation, so that for the same bang of precipitation you get a lot less buck of storable water. We're beginning to experience the pressures of climate change in the environment on our water resources. And the climate change models forecast reductions in surface stream flow of 10 percent to 30 percent relative to a 1900 to 1970 baseline.

So we are already in a deficit position and our income is dropping 10 percent to 30 percent. If this were a bank, we would declare it insolvent. The regulators would need to come in and take it over. That's what we might see on the Colorado. The Río Grande is maybe not in as extreme a condition, but it's still heavily allocated, probably overallocated, and it is probably even more vulnerable to climate change because its watershed does not reach as far north as the Colorado's, into regions where precipitation is more likely to increase. There are big questions about how much water various entities like the Middle Río Grande Conservancy District actually use and how much they're actually entitled to. Astonishingly enough, even after the better part of a century, there are still big, big ambiguities about who owns what and who's entitled to what. Indian water rights on both rivers remain big question marks.

If we were a really well-organized, foresightful society, we would clean up all this business as we approach the future. But we don't seem to be doing that. I liken climate change to a set of rapids on a river. Here we are in the boat of society and we're going to run the equivalent of Lava Falls in the Grand Canyon. Well, you can't control a lot of things down there in the rapids, but you can control one thing, and that is your point of entry. It's the location, the angle of your boat, your velocity. Point, angle, velocity. You can control those things. That's how you run a rapid. You've got to get the right point of entry, the right angle, the right speed, et cetera. Are we doing that? No, it doesn't look like it. So we're going to have a much more haphazard run through the rapids. We're going to be bouncing off the rocks, bouncing off the walls. We're going to be taking on a lot of froth. We'll get turned around probably. God only knows how we're going to come out the other end.

JL: That's a great metaphor. You can also get sucked right down to the bottom.

WD: At Lava Falls Rapid, if you don't have the right point of entry you can go over the ledge and into the big hole and get maytagged to a fare-thee-well. Our society can do that too.

JL: One of the things I've been thinking about a lot that we're factoring into this project is the role of an economically dominated paradigm in relation to how do we steer the boat.

WD: Yes.

JL: Obviously economics is a major factor, but it has to be seen, at least in my mind, as just one of a system of factors rather than the prevailing, predominant factor. What are your thoughts about that?

WD: The way I come to the economics issue is really through reflections on water conservation. For decades now our utilities and our leaders have been telling us that in the dry Southwest the key to water sufficiency is conservation. I don't believe that's true. I believe water conservation is a hoax. Let me explain why. If you or I use less water—we take short showers, xeriscape the lawn, use low-flow toilets, all the things that really we should do that seem morally and ethically correct—we're not adding to the resilience of our watershed, of our community, of our water utility, wherever we live. We're just making more water available for the next housing development or strip mall down the road, because conservation, under our present economic paradigm, doesn't go back to the river or the aquifer, to the source of the water. It just goes to more economic growth.

Water conservation is a really good way to keep short-term building, real estate, and other economic activities rolling. But long-term, what it does is harden demand. The concept of demand hardening is a crucial one when we talk about water in the Southwest. In times of drought in the past, waste has been our best friend, because it's easy to tighten up the system. You stop washing the car. You stop watering the lawn. You don't make puddles of any kind. So in a time of drought, when there's a wasteful system you can cut those wastes, and demand drops like a stone, and you make it through the drought.

If everybody has conserved, then the uses to which you put water are increasingly, maybe all, absolutely essential uses. You can't stop doing them. So when drought comes, if a community has really conserved but then has reallocated that conserved water to new growth, you can't pare it back. There's no flex in the system. Demand has hardened. You're in a very tight corner then.

The solution, according to many of our utilities, is then to do augmentation, to bring in new water through desalination or tapping new aquifers, even interbasin transfers—all these things that have enormous fiscal and usually environmental costs, and that can keep the hamster wheel of increased need, conservation, and demand hardening spinning for another generation or so. Ultimately we've got these two lines, of available supply and need, and they ultimately cross at a certain point. We keep thinking we can put off that point of crossing into the indefinite future. But one day, the day of reckoning will come. And with climate change, what will happen is that day of reckoning accelerates back toward the present out of the future, because supply is going to be dropping.

What do we do? The only real solution is not conservation. It's a changed economic paradigm. We have to figure out a way to transition to a steady-state economy rather than a continuous-growth economy. This is true in water, it's true in energy use, it's true with regard to greenhouse gases, and it's true in so many areas. The sooner we get on with the business of making this transition, the less painful it's going to be. It's like that image of the rapids that we used a little while ago. If we don't prepare, if we don't select our point of entry, then we give ourselves up to dumb luck and chaos, and they're not always very friendly to us as we go through the rapids of the future.

JL: That's really well put. I think about this all the time with regard to corporate economics that dominates much of our paradigm. It's not so much them versus us but, rather, all of us participating in this economically dominated paradigm based on, as Ed Abbey said, growth for the sake of growth, which is the ideology of a cancer cell.

WD: Exactly.

JL: Somehow we have to come to terms with that. In the meantime, I take a long hard look at the ways of governance that we have created in this country and how that contributes a huge amount of energy to the current economically dominated paradigm. Modern governance goes hand in hand with forwarding growth for the sake of growth.

WD: Exactly. You can't get reelected if the GDP is falling. So you run the economic system as best you can to goose that GDP figure up. That's how our politics work.

JL: What we're forwarding in this project is governance from within the commons rather than from the top down. This is not to say to take out the federal government but to really increase the sense of responsibility from a population base within the commons and reorganize the system of laws that were erroneously wrought in the first place. It's really interesting looking at the 1922 Colorado River Compact and what that implied as necessary to maintain over the future. Of course it was also based on an erroneous concept of how much water was coming down the river. Could I ask you to talk about your sense of the commons itself and how we might be able to restore a sense of human balance?

WD: I think about that a great deal. We are citizens of so many commons all the time, in a sense. The atmosphere, which we're polluting with greenhouse gases, is a commons. Our watersheds are a commons. Qualities like silence are a commons. These days in America you can hardly go any public place without hearing a doggone TV yammering away. Commons is what we share with each other, and somehow linked to this idea of transitioning to a steady-state economy we've got to transition to a set of values that is rooted in what we share together, rooted in a sense of community, rooted in our bonds to each other, rather than in our individual existences to the degree that we are so preoccupied today. Where that leads is to an emphasis on quality of life rather than on quantity of consumption.

Bill McKibben has done some very interesting work and has tracked a number of interesting trends and polling systems and so forth, and he is fond of citing the fact that it seems that as Americans accumulated more possessions, they report themselves as being less happy. More toys don't mean more fun. I think as we take this consumer society to its accumulating, trash-producing *reductio ad absurdem*, we're finding that it's not a very satisfactory way to live. Somewhere down the line, if we're going to be successful, if we're going to get into the future more or less intact, we've got to reorient our values to the primacy of the commons, to a sense of loyalty to place, to a sense of dedication of quality of life rather than just to the continuous consumption of more and more material things.

JL: Right on. Something that I think about a lot is that in a sense, the commons in which we participate the most is the commons of human consciousness, what Western culture has wrought on human consciousness. The whole consumer notion with regard to how we comport ourselves has splayed throughout human consciousness. To me the huge challenge is how to convey to the mainstream commons of human consciousness what really needs to happen. That's why I really love talking to peoples of indigenous persuasions, because they still retain systems of values that provide far more balance with habitat.

WD: Most people through most of human history have lived in something like a steady-state economy. This continuous-growth way of life that we define as normal today is really an artifact of the modern era in western Europe, then exported to North America, and really to the rest of the world today. But that's not where human beings evolved. That's not where we lived for all those millennia beforehand. We can do this, and that's the wisdom that we get from Native peoples, because they're still connected to that other reality, that longer heritage of human experience. They can draw on traditional knowledge about how that steady-state world existed and what kinds of values were necessary to underpin it.

JL: And that's why their perspectives remain absolutely vital in today's world. Thanks, Bill.

Conclusion

An unfortunate fundament of American law is to legislate in order to serve the wants and perceived needs of those who would privatize common natural resources for personal gain. This sets a cultural standard that apparently works until the population outgrows the carrying capacity of the common landscape. At this point, national legislation and natural law fail to coincide, and the gap between them gradually widens exponentially until the system crashes, and the population diminishes by virtue of overexploitation of common resources. This should be self-evident, especially as we look at our world's repertoire of disaster potential that increases daily.

Let us recall the five exponentially growing issues forwarded by the *Limits to Growth* project mentioned in the introduction:

population
food production
industrialization
pollution
consumption of nonrenewable resources

And we add a few more issues of growing probability:

world war
pandemic
economic collapse
cybercollapse
climate instability and global warming

Obviously we are in the midst of a complex system of factors of our own creation that poses increasing peril to the entire biotic community including our own species. In this world, the territorial imperative is enacted within a mosaic of nationalistic phenomena such as we see today, each nation complete with its own legislated laws and system of standards, all vying for position within the prevailing collective economic house of cards. It certainly doesn't take a particularly gifted mind to extrapolate the inevitability of disaster. Rather, it requires some strength of character to look directly into the eye of darkening collective human consciousness for any spark of insight, any glimmer of understanding as to how we attempt to wend our way back from the edge of the abyss.

I do not find hope and light in the seats of centralized political power held in sway by corporate economics anywhere in this world. Rather, the great hope that I find lies in rural communities attuned to respective homeland where self-sustainability has been traditionally maintained for long enough to become a resilient culture of practice. I also find hope in those cultures of practice where a spiritual relationship exists with homeland such as we see in the indigenous and traditional communities reflected in the earlier essays in this book. I find hope in the restoration of sustainability and the reawakening of intuitive understanding of our place as a species within the flow of Nature. I find hope in decentralized governance from within naturally defined geographical areas—namely watersheds. And I find tentative hope within an enlightened scientific community intent on evolving technology that is applicable to solving many within the array of areas of jeopardy that we face as a planetwide biotic community.

The great anarchist philosopher and geographer Pyotr Kropotkin defined his sense of an ideal society in his essay "Modern Science and Anarchism," published in French in 1913.

> The anarchists conceive a society in which all the mutual relations of its members are regulated, not by laws, not by authorities, whether self-imposed or elected, but by mutual agreements between the members of that society and by a sum of social customs and habits—not petrified by law, routine or superstition, but continually developing and continually readjusted in accordance with the ever-growing requirements of a free life stimulated by the progress of science, invention and higher ideals.
>
> No ruling authorities, then. No government of man by man; no crystallization and immobility, but a continual evolution—such as we see in nature.[1]

In 1985, my friend Gary Snyder spoke his carefully thought-out definition of bioregionalism into my microphone:

> *Bioregion*, the term itself, would refer to a region that is defined in some way by its plant and animal characteristics, its life zone characteristics that flow from soil and climate—the territory of the Douglas fir or the region of coastal redwoods; short-grass prairie, medium-grass prairie, and tall-grass prairie; high desert and low desert. Those could be, or verge on, bioregional definitions. When you get it more specific, you might say Northern Plains short-grass prairie, Upper Missouri Watershed, or some specific watershed of the Upper Missouri. The criteria are flexible, but even though the boundaries and the delineations can vary according to your criteria, there is roughly something we all agree on. Just like we agree on what a given language is, even though languages are fluid in their dialects. So bioregionalism is a kind of creative branch of the environmental movement that strives to reachieve indigeneity, reachieve aboriginality, by learning about the place and what really goes on there.
>
> Bioregionalism goes beyond simple geography or biology by its cultural concern, its human concern. It is to know not only the plants and animals of a place, but also the cultural information of how people live there—the ones who know how to do it. Knowing the deeper mythic, spiritual, archetypal implications of a fir, or a coyote, or a blue jay might be to know from both inside and outside what the total implications of a place are. So it becomes not only a study of place, but a study of psyche in place. That's what makes it so interesting. In a way, it seems to me that it's the first truly concrete step that has been taken since Kropotkin in stating how we decentralize ourselves after the twentieth century.[2]

Many of us perceive John Wesley Powell as a man who attempted to put bioregionalism into practice even though the term was coined over six decades after Powell's death. His rendering of the map of Watersheds of the Arid West that appears in the *Eleventh Annual Report to the United States Geographic Survey 1889–90* can well be regarded as a map of the bioregions of the United States west of the 100th meridian.

Powell recognized watersheds as commons and clearly defined this in the *Eleventh Annual Report to the USGS*. Many modern conservationists and environmentalists are hearkening to Powell's early wisdom.

The essays that appear in this book were consciously written or spoken from within the home watersheds of each of the authors. Each essay is a presentation of the workings of a different cultural mind within the context of respective homeland. Each of the authors has both practical and spiritual ties to the North American Southwest that comes from dipping deeply into the flow of Nature, of attuning the senses to the smell of juniper, the sound of the coyote, the sight of endless space, the taste of dust, the feel of the wind—the deep intuition of the sense of being kindred with the life forms, the landscape, the sky, the living waters.

It is not superstition to intuit and celebrate the urge to life and consciousness that occurs because of the relationship of the Earth to the Sun. Rather, it is part of humankind's endowment to be granted the capacity to perceive the sacred quality of homeland, to feel awe at the great mystery that pervades the universe—and to be humbled by the recognition that our species is but the tiniest part of that great mystery, that we are privileged to possess a level of consciousness that far exceeds our untimely paltry pursuits that have brought us to the edge of oblivion.

At the beginning of the twentieth century, roughly 40 percent of Americans lived in cities. A century later, nearly 80 percent live in cities, leaving 20 percent of us in rural America. The toll of cities on surrounding habitat is horrific, but consciousness of this toll is meager at best. We as a culture are into several generations' worth of major overindulgence. Over the last generation, much of our attention has been dominated by digital technology to the extent that many of us spend precious waking hours immersed in a two-dimensional virtual reality that lures our collective and individual attention away from the real McCoy, albeit we have created a mighty force for instant communication throughout our digitally attuned culture of practice.

There is a large cadre of scientists of various persuasions monitoring the health of the environment, for which I am profoundly grateful. They also vigorously warn us that we are face to face with enormous climate change due to global warming that is greatly affecting our planet's atmosphere. But too few of us now live out in it enough to hear the song of homeland, to see its glow, to register its ambience, to take the measure of its pulse—to dance with the spirit of place. There *is* a beam of hope. A small but growing coterie are turning back to homeland, to home watershed, to home foodshed, and are consciously working hard in behalf of sustainability and rebalance within the flow of Nature. Many now listen to our indigenous neighbors for insight into technique and, even more important, state of mind.

In today's world of economic/political culture, there exists an immense polarity between the right and the left. Unfortunately, they are mostly to the right and left of the wrong issues. A hundred years ago Pyotr Kropotkin campaigned vigorously in behalf of peasants whose lives were usurped to fulfill the presumed needs of the governing gentry. Kropotkin wanted to educate these peasants so that they could take their rightful place within the cultural continuum. At that time our human population of the planet was less than 25 percent of today's population. There yet remain disgraceful disparities between the "classes" worldwide, and the polarizations grow ever more vicious. But these polarizations blindside us to the single most important issue at hand—the health of our planetary habitat. Our consciousness is so dimmed by our obsession with economics and the deteriorating human condition that we cannot accept the fact that growth for its own sake is inevitably a suicidal course. The longer this course prevails, the less likelihood we'll survive as a species. We have allowed distant governance to grow too great. The collective consciousness of our governing bodies is far too narrow and modest for the task at hand. Thus, those of us with the means luxuriate in the final days of our perceived overabundance.

Techno-industrial-military culture and its attendant political complex has grown too enormous and inflexible to find the resilience necessary to survive itself, let alone the coming decades of global warming and climate instability. The economic imperative has become such a dominant force in American culture and beyond that our federal government's attention to the health of our overall biotic community has been basically put on hold. Thus, gradual decentralization of political power is vital over the near future so that governance from within specific habitats and watersheds takes priority over centralized governance that is too casually influenced by the moneyed oligarchy that has come to prevail in America and beyond. Money has paid for legislation that is counter to natural law; thus our wildlands dwindle and biodiversity withers in the wake of our species' march of progress in an erroneous quest of never-ending growth.

It requires vigorous and relentless will and deep courage from the grassroots to stay this juggernaut to both achieve a steady-state, no-growth economy and concurrently protect habitat from extractors and developers who see habitat solely as a source of income. Decentralization of political power is indeed a revolutionary act, yet if it is founded in a land ethic born of conscience and conscious concern for the health of habitat, it is right and true even if it breaks laws designed to define and defend procedure legislated in behalf of extraction and development of homeland for financial profit. It is obvious that centralized governance funded by corporate economics is blindsided to the level of jeopardy to the biotic community that we as a species have collectively wrought during our tenure as the keystone species.

Thus, we must restore at least some measure of decentralized self-governance from within an appropriate perceptible span of landscape. And what could possibly be more obvious than home watershed? To take it a step further, the home watershed/foodshed should preside at "the head of the table" of the governing body that determines carrying capacities, interspecies cooperation, its own health, and myriad other factors necessary for maintaining a state of honest balance.

How to do this? Starting at the local level is the only way to invigorate change. Culture evolves from the grassroots. A good way to begin is to organize a celebration of home watershed with poetry, music, and artful reflections on the beauty of homeland—especially at times of planting and harvest. The vernal and autumnal equinoxes are among the most special times of year, times when the incense of local plants should fill the air, times when processions of people should pass across the face of the land in joy, introspection, and reverence as we pay homage to the spirit of place. These are times to desecularize homeland, sacrifice economics, and reinvoke the sense of the sacred quality of the watershed; elevate consciousness and dance with the local deities; marvel at the miracle of existence; and reflect on how fragile and tenuous life is in this enormous span of universe. And then maintain this heightened state of mind and consciousness from equinox to equinox, solstice to solstice.

It is good to recognize that farmers markets and community co-ops are already existing centers of cultures of practice of sustainability. It is from here that the handcrafted lifestyle gains impetus and finds outlet. Here people who are already working within watershed/foodshed consciousness prevail, and it is here that organizing governance from within home watershed may begin to formulate, but without heavy bureaucracy. Giving creativity its due is essential to creating and maintaining practices appropriate to home watershed. I can imagine watershed yoga; seminars in ecology; conscious evolution and practice of foodshed culture; ballads of homeland; river trips; annual homeland clean-up brigades; seminars in appropriate economics; honoring of elders; restructuring within the purview of ever-evolving basic minimum technology and its appropriate distribution; watershed watch; schoolboard participation in "ecosystemology" for K–12; cadres of scientists presenting characteristics of regional watersheds that need monitoring including groundwater depletion and recharge, areas of pollution, anticipated annual snowmelt, assessments of carrying capacities, general health and balance of watershed, appropriate energy production sources, and as many other appropriate factors as can be imagined within the scientific purview; seminars in mutual cooperation and reinventing appropriate cultural relationship to homeland; defusing commoditization of common resources; recording and documenting stories of homeland; evolving appropriate participatory traditions—festival of the waters, festival of the birds, festival of the harvest, festival of the elders, festival of the newly born. In other words, invigorate cultural realignment with homeland. Participate democratically with more emphasis on home watershed/foodshed than national politics; gradually soften federal governance and defuse corporate control; gradually reduce human population; create local energy generation sources using nondestructive basic minimum technology; set goals of restoring our planet to a state of balance—truly become the mind of the planet. This is a utopian ideal that hearkens back to a meld of Kropotkin's concept of society, Snyder's definition of bioregionalism, indigenous peoples' sense of kinship with all species, and appropriate application of ever-refining science, all within a fomenting, evolving social consciousness within home bioregion and beyond. Indeed, the planet Earth is our home bioregion. No other heavenly body will ever be so hospitable.

It is important to recognize the degree to which biodiversity and cultural diversity are interconnected. Cognitive diversity is essential for responding to crisis. We face major crises that could well take us and many other species down. We have wrought enormous damage to our planetary environment; we have but a limited and very dangerous and selfish monocultural point of view. We have to look through many windows of consciousness, many points of view to find our way home again. Our true goal must be to reestablish our collective sense of indigeneity to this exquisitely beautiful and bountiful homeland of our planet and to proceed mindfully within the realization that Nature's bounty is a common resource for all living creatures past, present, and future. And that no species has any right whatsoever to privatize these common resources, these common waters.

Traditional indigenous cultures, and other cultures that have evolved long-term sustainability within specific habitats, retain the adaptability to shifting conditions including resilience within the inevitability of change. The Native cultures of the North American Southwest still retain strong traditional practices that align themselves with the flow of Nature. Their sense of place is enduring, and they have much to teach the rest of us about recognition of our kinship with fellow species, rootedness within the biotic community, and the sacred nature of life, consciousness, and homeland. In the most real sense, the indigenous nations and communities of the North American Southwest are "seedpods" of human survival potential over the coming century. We should heed their wisdom and learn to respect their intuitive understanding of relationship of culture to homeland.

Indeed, people of the land point the way to the practice of restoration ecology in home habitat and, concurrently, our cultural restoration as we reengage with the land. To desecularize the landscape is a profound concept that we would do well to enshrine as the epitome of a land ethic. Tithing habitats for protection within each ecosystem might ensure survival of biodiversity as we understand it. To resurrect handcrafted lifestyles as revealed in many of the chapters of this anthology may lead the way out of the spiritual lethargy of consumerism into lives filled with meaning.

Consciousness is Nature's enormous gift to our species as we have evolved over scores of millennia. How we use this gift of consciousness will determine how our role either develops or decays over the coming decades. Reveling in the flow of Nature leads to a level of spiritual awareness available nowhere else. Fully engaging the five bodily senses in wildlands is to reawaken lost sensibilities and redefine perspective. Looking deep into the heart of our galaxy of a summer nighttime while listening to coyotes sing allows a glimpse into just how profound the gift of consciousness is.

There is tremendous consciousness to be gleaned from deep perception of home watershed. To think like a watershed requires the evolved capacity to perceive interconnectedness in motion. To understand the watershed/foodshed as commons requires the consciousness of the absolute egalitarian.

In consciousness we trust.

—Jack Loeffler

Notes

1. Rodger N. Baldwin, ed., *Kropotkin's Revolutionary Pamphlets* (New York: Dover, 1970), 157.
2. Jack Loeffler, *Headed Upstream: Interviews with Iconoclasts* (1989; repr. Santa Fe, NM: Sunstone Press, 2010), 180–81.